POWER SUPPLY TESTING HANDBOOK

Join Us on the Internet

WWW: http://www.thomson.com
EMAIL: findit@kiosk.thomson.com

thomson.com is the on-line portal for the products, services and resources available from International Thomson Publishing (ITP).

This Internet kiosk gives users immediate access to more than 34 ITP publishers and over 20,000 products. Through *thomson.com* Internet users can search catalogs, examine subject-specific resource centers and subscribe to electronic discussion lists. You can purchase ITP products from your local bookseller, or directly through *thomson.com*.

Visit Chapman & Hall's Internet Resource Center for information on our new publications,
links to useful sites on the World Wide Web and an opportunity to join our e-mail mailing list.
Point your browser to: **http://www.chaphall.com** or
http://www.thomson.com/chaphall/electeng.html for Electrical Engineering

A service of

POWER SUPPLY TESTING HANDBOOK

Strategic Approaches in Test Cost Reduction

Earl Crandall

Southern East-Coast Field Applications Engineer
Micro-Linear Corporation
Raleigh, NC

CHAPMAN & HALL

I(T)P® INTERNATIONAL THOMSON PUBLISHING

New York • Albany • Bonn • Boston • Cincinnati • Detroit • London
Madrid • Melbourne • Mexico City • Pacific Grove • Paris • San Francisco
Singapore • Tokyo • Toronto • Washington

Cover design: Curtis Tow Graphics

Printed in the United States of America

Chapman & Hall
115 Fifth Avenue
New York, NY 10003

Chapman & Hall
2-6 Boundary Row
London SE1 8HN
England

Thomas Nelson Australia
102 Dodds Street
South Melbourne, 3205
Victoria, Australia

Chapman & Hall GmbH
Postfach 100 263
D-69442 Weinheim
Germany

International Thomson Editores
Campos Eliseos 385, Piso 7
Col. Polanco
11560 Mexico D.F
Mexico

International Thomson Publishing–Japan
Hirakawacho-cho Kyowa Building, 3F
1-2-1 Hirakawacho-cho
Chiyoda-ku, 102 Tokyo
Japan

International Thomson Publishing Asia
221 Henderson Road #05-10
Henderson Building
Singapore 0315

1 2 3 4 5 6 7 8 9 10 XXX 01 00 99 98 97

Library of Congress Cataloging-in-Publication Data

Crandall, Earl.
  Power supply testing handbook : strategic approaches in test cost reduction / by Earl Crandall.
    p.   cm.
  Includes index.
  ISBN 0-412-08841-X (alk. paper)
  1. Electric power supplies to apparatus -- Testing. I. Title.
TK7868.P6C73   1997
621.31'7--dc20   96-42320
                                                          CIP

British Library Cataloguing in Publication Data available

"Power Supply Testing Handbook" is intended to present technically accurate and authoritative information from highly regarded sources. The publisher, editors, authors, advisors, and contributors have made every reasonable effort to ensure the accuracy of the information, but cannot assume responsibility for the accuracy of all information, or for the consequences of its use.

To order this or any other Chapman & Hall book, please contact **International Thomson Publishing, 7625 Empire Drive, Florence, KY 41042.** Phone: (606) 525-6600 or 1-800-842-3636. Fax: (606) 525-7778. e-mail: order@chaphall.com.

For a complete listing of Chapman & Hall titles, send your request to **Chapman & Hall, Dept. BC, 115 Fifth Avenue, New York, NY 10003.**

Dedication

I want to thank my family; my wife, Grace, and my sons, Tim, Tom and Ted; for their love and understanding for the years during which I was writing this book. Their support, during this time when I was not as available as a husband and father should be, made this work possible.

I wish to thank Frank Sledd for his assistance in the early stages of this project. He provided input for several chapters.

Moreover, I want to publicly give thanks and praise to my Lord Jesus Christ for His unfailing love and the gift of salvation through His death on the cross. I thank Him for the abilities He has given me to develop and use.

Table of Contents

Introduction

The power electronics industry has been growing at a rapidly increasing rate, but analog and power supply testing have not really kept pace. The digital test industry has been growing at leaps and bounds, and now the analog and power supply industry is starting to take off. This book is intended to be a catalyst for the thought process occurring during all phases of the creation of Power Supply test strategies, power supply test processes, and the power supply tests themselves. May it make your task a more joyous occasion.

This book is generally separated into three sections, covering the subject of power supply testing from the aspects of the design engineer, the manufacturing engineer, and the field service engineer. These lines were arbitrarily drawn and should be crossed by every reader, for each impacts the other. In fact, you should find that organization of your business is driven by the needs of the marketplace, your location, your company size, your product mix, and the many personalities involved in the operation. Use these differences as stepping stones to enhance your future success.

A nongoal of testing a product is the repair of that product for shipment, although we all do just that. The real purpose of testing and diagnosing a power supply is to determine just exactly where the manufacturing process went wrong, and then fix it so that "bad" products are not manufactured. Any other goal is just a waste of good time and money. For instance, if you have a fresh lot yield of 60%, you will have 40% failures. A 40% failure rate is a failure rate of 400,000 ppm (parts per million). It takes a sizable test and repair operation to repair 400,000 power supplies (out of every million you manufacture). When the goal of repair is to drive faults back to the point of origin and fix the source of the process error, then fresh lot yields will increase until failure rates are very small. This **WILL** reduce manufacturing costs and **INCREASE PROFITS!** So with this goal in mind, read on!

Safety Information

It is important to note that test procedures, techniques, and tools used for testing power sources, as well as the skill and experience of the individual performing the work vary widely. It is not possible to anticipate all of the conceivable ways or conditions under which power sources may be tested, or to provide cautions as to all of the possible hazards that may result. Standard and accepted safety precautions and equipment should be used when handling high voltages, high currents, and high-frequency energy.

Some procedures require the use of tools (or pieces of equipment) specially designed for a specific purpose. Before substituting another tool (or piece of equipment), you must be completely satisfied that neither your personal safety, the performance of the unit under test, nor the validity of the test results will be endangered.

Although information in this handbook is based on industry sources and is as complete and accurate as possible at the current revision level, the possibility exists that changes will occur in the future to compensate for: changing technologies, adopted changes, omissions, and errors. While the author has striven for total accuracy, responsibility cannot be assumed for any errors, changes, or omissions that may occur in the compilation of this handbook.

Proper test procedures are vital to the safe, reliable operation of all power sources, as well as the personal safety of those performing tests and repairs. This handbook outlines processes and procedures for testing power sources using safe, effective methods. This handbook contains many **NOTES, CAU-TIONS,** and **WARNINGS** which should be followed along with standard safety procedures to eliminate the possibility of personal injury or improper conditions that could damage the device under test or compromise its safety.

Part or Model Numbers

Any part numbers or model numbers listed in this reference are not recommendations by the author or publisher for any product by brand name. They are merely references that can be used for comparison purposes to locate each brand supplier's discrete part number.

Definition of Terms

CAUTION

Statements of **CAUTION** identifying conditions or practices that could result to damage to equipment or property.

WARNING

Statements of **WARNING** indicate a personal injury hazard may exist.

DANGER

Statements of **DANGER** identify conditions or practices that could result in personal injury or loss of life. Additionally, damage to equipment or property may result.

DANGER

ALL PARTS OF A POWER SUPPLY ASSEMBLY INCLUDING THE INPUT CIRCUIT COMMON MAY BE AT OR ABOVE THE POWER LINE VOLTAGE. THE ENERGY AVAILABLE AT ANY POINT ON A POWER SUPPLY ASSEMBLY MAY BE LIMITED ONLY BY THE FUSE OR CIRCUIT BREAKER PROTECTING THE CIRCUIT FROM WHICH IT IS POWERED (CONNECTED). DO NOT ATTEMPT SERVICE OPERATIONS WITHOUT THE CORRECT TRAINING, EQUIPMENT, AND DOCUMENTATION.

DANGER

DO NOT ATTEMPT OPERATION OF ANY POWER SUPPLY WITHOUT FIRST READING THE INSTRUCTION MANUAL.

DANGER

ELECTRICAL SHOCK HAZARD EXISTS WHEN A POWER SUPPLY IS CONNECTED TO A TEST INSTRUMENT. DO NOT TOUCH, CONNECT, OR DISCONNECT ANY CONNECTIONS OR CABLES WHILE THE POWER IS APPLIED.

DANGER

REMOVE ALL WATCHES, RINGS, AND JEWELRY BEFORE WORKING AROUND ANY ELECTRICAL EQUIPMENT OR BEFORE PERFORMING ANY TESTING.

DANGER

DO NOT WORK IN ANY AREA WHERE THE FLOOR IS CONDUCTIVE WITHOUT SPECIAL EQUIPMENT. THIS INCLUDES ANY CONCRETE, METAL, FLOORS WITH CONDUCTIVE TILE, OR WET OR DAMP FLOORS.

WARNING

TO PREVENT POSSIBLE ELECTRICAL SHOCK OR DAMAGE TO EQUIPMENT, CHECK LOCAL ELECTRICAL STANDARDS BEFORE SELECTING A POWER CORD OR MAKING VOLTAGE SELECTION. THE INFORMATION PRESENTED IN THIS HANDBOOK MAY NOT BE CORRECT FOR ALL LOCATIONS. ALWAYS CONSULT LOCAL STANDARDS.

WARNING

REMOVAL OF INSTRUMENT COVERS MAY CONSTITUTE AN ELECTRICAL HAZARD AND SHOULD BE ACCOMPLISHED BY QUALIFIED PERSONNEL ONLY.

WARNING

DISCONNECT ALL POWER TO ANY EQUIPMENT BEFORE REPLACING ANY COMPONENTS, MODULES, OR SUBASSEMBLIES. FAILURE TO DO SO MAY RESULT IN ELECTRICAL SHOCK.

WARNING

TO AVOID ELECTRIC SHOCK FROM DANGEROUSLY HIGH VOLTAGES, FOLLOW THE APPROPRIATE EQUIPMENT MANUAL INSTRUCTIONS AND PROCEDURES WHEN WORKING WITH HIGH-VOLTAGE POWER SUPPLIES.

CAUTION

BECAUSE OF DIFFERING POWER REQUIREMENTS, POWER SUPPLIES MAY REQUIRE A DIFFERENT POWER CORD CONNECTOR. WHEN PLACING A NEW CONNECTOR ON THE POWER CORD, CARE MUST BE TAKEN TO ENSURE THAT ALL THREE WIRES ARE CONNECTED PROPERLY. THE GREEN OR GREEN/YELLOW-STRIPE WIRE IS ALWAYS CONNECTED TO EARTH GROUND (E). THE WHITE OR LIGHT-BLUE WIRE IS CONNECTED TO THE NEUTRAL SIDE OF THE POWER LINE (N). THE BLACK OR BROWN WIRE IS CONNECTED TO THE HIGH SIDE OF THE POWER LINE (L). FOR EQUIPMENT DESIGNED FOR USE IN OTHER COUNTRIES, CONSULT LOCAL STANDARDS.

CAUTION

WHEN PERFORMING ANY CALIBRATION OR MAINTENANCE OPERATIONS, DO NOT REMOVE OR REPLACE ANY CIRCUIT CARDS OR ASSEMBLIES WHILE THE POWER IS TURNED ON. FAILURE TO TURN POWER OFF MAY RESULT IN ELECTRIC SHOCK OR DAMAGE TO THE EQUIPMENT.

CAUTION

AVOID THE USE OF CHEMICAL CLEANING AGENTS THAT MIGHT DAMAGE ANY PLASTICS OR INSULATIONS USED IN POWER EQUIPMENT. DO NOT APPLY ANY SOLVENT CONTAINING KETONES, ESTERS, OR HALOGENATED HYDROCARBONS WITHOUT FIRST CONSULTING MATERIAL SAFETY DATA INFORMATION, MANUFACTURER'S SPECIFICATIONS, INSTRUCTION MANUALS, AND TAKING ALL APPROPRIATE SAFETY MEASURES FOR THE USE OF SUCH CHEMICALS.

CAUTION

EXAMINE ALL EQUIPMENT AND MANUALS FOR CAUTIONS, WARNINGS AND DANGER STATEMENTS, WARNINGS OR LABELS AS A FIRST STEP PRIOR TO PERFORMING ANY DISASSEMBLY, TEST, REPAIR, OR OPERATION.

CAUTION

ALWAYS KEEP ALL APPLICABLE MANUALS ACCESSIBLE TO THOSE PERSONS INVOLVED IN THE USE, MAINTENANCE, OR REPAIR OF POWER PRODUCTS.

PREPAREDNESS and PLANNING

IN ALL LOCATIONS WHERE ANY HAZARD EXISTS DURING THE OPERATION, TEST, REPAIR, OR CALIBRATION OF ANY POWER SUPPLY, POST COPIES OF THE CORRECT PROCEDURES TO FOLLOW, IN THE EVENT OF ANY INJURY TO ANY PERSON, INCLUDING AS A SUGGESTED MINIMUM, THE FOLLOWING:

- A POSTED LIST OF TRAINED FIRST RESPONDERS AND CONTACT NUMBERS.
- A POSTED LIST OF EMERGENCY PHONE NUMBERS AND PERSONNEL.
- A POSTED LIST OF EMERGENCY INSTRUCTIONS.
- A POSTED LIST OF AUTHORIZED PERSONNEL TO ENTER HAZARDOUS AREAS.
- A POSTED LIST OF AUTHORIZED AND TRAINED PERSONNEL TO PERFORM HAZARDOUS FUNCTIONS.

1

Product Specifications

The one purpose of test following design is to verify product specifications, sometimes called design verification testing (DVT). In this chapter we will examine product specifications that are the root of the tests that will follow. This does not mean that each of these specifications must be tested in volume production. Many specifications, once checked, tested, or verified on the first few pieces by engineering, are ignored by the manufacturing test process. Some companies do verify these specifications in a sample from production as a part of an ongoing reliability program. The first task in looking at testing a product is to review the specifications very carefully. It is the test against product specifications that determine whether or not the project has been a success. The tests are your yardstick. Not all specifications are written to the same level of detail. Caution must be exercised when dealing with a loosely written specification. Any additional input from the customer must be documented for the design specification that you may generate and use internally. Some customers may even provide typical load hardware to test the power supply design against. This is not the best way to approach the solution; where possible, customers should be encouraged to reach a specification for the product to define and document just what it is. Over time, the customer's product and load requirements may change, and without a specification to modify, documentation of any change is not only difficult, product control will be nearly impossible.

Physical Specifications

The physical specifications of a power product are not usually tested once a design leaves engineering, unless there is a specific requirement to do so. Nonetheless, this is an important area that is often overlooked. Some specifications are very vague about where the volume is to be filled. The application may require specific physical attributes not clearly pointed out in the specifications. It is always important to work with the customer and understand their application of your product thoroughly. Physical dimensions may be unusual in that the product may be required in an unusual shape. On some occasions the size may actually be critical, and testing of the exact size, or some of the dimensions may be required. Jigs and fixtures may help when this is a problem. Usually careful attention to the design of the package will prevent problems from occurring in this area. Where tolerances are close, some consideration may need to be given to the thermal expansion of some materials.

Mounting Requirements

One area that needs particular attention is the specifications for mounting the power supply. The size and location of mounting holes and hardware must be specified. A drawing or blueprint will usually be referenced, see Figure 1-1 for example. The original (customer's) specifications for mounting need to be copied to the specification or print, illustrating the mounting holes. It is important that reference points do not become transposed or swapped, which may allow tolerance buildup problems to creep into production in the future.

METRIFICATION

Remember to examine the specifications for metric dimensions and sizes. Many foreign countries, (including some U.S. vendors shipping to foreign customers) may require the use of the metric system for dimensioning as well as for hardware.

Volume

Volumetric tolerances can sometimes be very demanding. As a result, mechanical testing operations may sometimes be required to meet specifications. This can especially be the case for open-frame power supplies that are not constrained by a physical case or cover. It's always good to remember the following critical points, especially when volumetric constraints are very

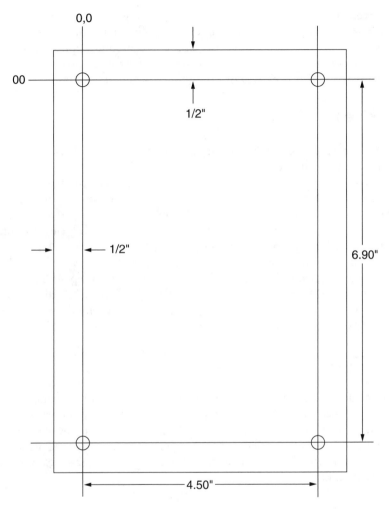

Figure 1-1. Product Mounting Location Specification

tight. They are the concepts of "Design for Testability", "Design for Manufacturability," and "Design for Repairability."

Weight

For some applications, weight can be a major consideration. Switching
power supplies operating at higher frequencies help to solve this problem.
Good attention to the packaging philosophy will also allow the reduction

of the weight of some power supplies. Weight is generally a consideration when the power supply is intended one of the following applications: portable use, aircraft use, spacecraft use, and some mobile applications. The weight will need to be checked on prototypes, first piece production, and following any major revisions of the product or its larger components.

Finishes

Various finishes are available on the market today including: bare metal, chromate, anodized aluminum, copper plated steel, paints, and plastic coatings, all of which are available in many colors. Standards exist for the various grades of finishes for most materials. Metal and plastic surfaces are often measured for the 'roughness' of the finish, usually in microinches. A check with the various industry organizations will help you locate the standards you need. For instance, the Society of Plastics Industries (SPI) produces a kit called "The Mold Finish Guide" Catalog Number AR-106, which is a standard for plastic finishes. This Kit costs $10.00, prepaid only. It is available from SPI, Literature Sales Department, PO Box 753, Waldorf, MD 20604, 800-541-0736 for credit card orders . Similar standards exist for the finishes of metal materials as well.

Labeling

Units, terminals, connectors and adjustments may all require labeling. There are many adhesive labels or direct silk screening that may be used. Some open-frame products will use only the silk screening on the printed circuit board (PCB) for labeling. Where critical requirements exist, testing of finishes and labels are usually best left to testing services or laboratories with the proper capabilities.

Materials

Some applications may require the use of special materials or the exclusion of certain materials. Some applications or markets are sensitive to some materials. One example of this is the exclusion of cadmium plated parts and materials in products destined for shipment to Sweden and other countries.

Magnetic Fields

Components or circuits that may be sensitive to magnetic fields might be present in the area where the product is to be located, thus requiring special considerations. Careful consideration must be given to the specifications in this area. Generally it is a good idea to obtain a working sample of the product for which the power supply is being designed so that the reasons for any possible interactions may be better understood. In some applications, stray fields are not tolerated outside the power supply as they may interfere with magnetically sensitive components such as deflection systems on cathode ray tube (CRT) displays. "Mu metal" shields may be required for either the power supply, the CRT, or for any sensitive components in the vicinity. Testing for stray magnetic fields is usually accomplished in the intended product. In some critical applications, special testing may be required by a testing service if the capability does not exist in-house. This can be especially true for terminals and workstations or other systems containing some form of CRT.

Thermal

Generally, tests are made at specified operating ambient temperatures, both minimum and maximum values, as well as at a nominal room temperature of 25°C. For many commercial power supplies this temperature range may be from 0 to 50°C. Additionally, a storage temperature may be specified, typically −55°C to +70°C. The type of cooling as well as the direction of the airflow are usually stated where they are important. Forced water cooling is sometimes specified on very large power supplies. Other cooling techniques include freon baths, oil immersion, and pressurized gas or airless environments. Specifications may call for operation at temperatures other than at those suggested above.

Whatever range of temperatures is specified, the equipment may have to be tested during development. Occasionally, equipment will be specified for operation beyond specifications without damage. Equipment will not be expected to meet output specifications, but will be required to "not suffer damage" during this time. An example of this is operation following exposure to extremely low temperature during transport where specifications call for no damage when input power is applied until the unit under test reaches the specified operating temperature. While ambient temperature may be within limits, the internal temperature of the unit under test will be

Table 1-1. Typical Temperature Ranges

	Commercial	Industrial	Mil/Aerospace
Ambient Temperature Range	0 to +70°C	−25 to +85°C	−55 to +125°C

at a much lower specified temperature. Testing at specified upper temperature limits is highly recommended to determine junction and component temperatures. Testing at the specified lower temperature limit by cold soaking is also highly recommended to verify that the product will not damage itself under these conditions. These tests will also aid in the verification and prediction of mean time between failure (MTBF) figures.

Environmental

Some units will be required to operate in hostile environments and it may be appropriate to test these units under either actual or simulated environmental conditions. These conditions may include salt spray, high humidity, high altitudes, and corrosive atmospheres. Sometimes several of these conditions are present with extremes of temperature that require great care to be taken in the thermal design and packaging design of these products. For special testing of this nature, a certified testing laboratory may be contracted that specializes in the kind of testing needed.

Shock and Vibration

Some power supplies will require shock and vibration testing. This is especially valid where the end product will itself subject to shock and vibration. The power supply specifications will usually direct you the type of shock and vibration expected, as well as the type and amount of testing that is required. Again, testing services may be employed for this task, especially where the proper test equipment is not available.

Military Requirements

Military specifications will usually specify with great detail the type of testing that is to be performed, under what conditions, with what equipment and by whom it will be performed. It is typical of the military to write a complete set of requirements and specifications for each program, to satisfy the needs of that program. Careful consideration of program requirements and specifications to ascertain testing requirements is necessary.

Connectors

Input and output connectors are both electrical and mechanical in nature. Care must be exercised when specifying connectors from more than one vendor. Connector positioning can also be critical in some interconnection schemes.

Electrical Specifications

All electrical specifications must be tested for the purpose of design verification. This testing is usually not performed during the manufacturing process except where significant design changes have been introduced to the product. The electrical specifications of a power product are not all tested in the production environment once they leave engineering, unless there is a specific requirement to do so. This is because the main purpose of manufacturing testing is to detect process faults that have occurred during the process so that the process can be adjusted or modified to prevent the creation of that defect.

For custom power supplies, some applications may require specific electrical attributes not clearly pointed out in the specifications. For this reason, it is always important to work with the customer and understand his or her application of your product thoroughly. This may require that you perform some testing of prototypes in the actual application or a simulation of the application.

Input Specifications

Input tests are performed on the various inputs to power supplies. Most often the power inputs are alternating current (AC), and these are sometimes referred to as "AC to DC converters." Some power supplies are operated from a DC source such as a battery. These are usually referred to as "DC to DC converters." Other inputs may include bias voltages, clock signals, and control signals.

AC Input

This is usually the main source of power for the power supply. Alternating current Input voltages are generally specified to be measured at the unit under test, unless otherwise indicated by the specifications. The input voltage will be specified with a tolerance. The tolerance range may be different for the various topologies. Typically switchers are designed with the widest possible input tolerance range.

AC input impedance is a specification seldom seen in the past, but one to be looked for in the future. Complex systems and power requirements demand that this specification be available to the system designer.

In the United States, power companies typically maintain the voltage at the nominal value of 120 Vac. The nominal 120 Vac value may vary between minus 13.3 and +5.8%. This voltage range is considered the office machine range in which most home and office electronic equipment is designed to operate.

There are exceptions for some equipment such as large automated data processing (ADP) installations.[1,2] A useful publication is *A Guideline on Electrical Power for ADP Installations,* Federal Information Processing Standards (FIPS) publication 94, dated September 21, 1983, published by the U.S. Department of Commerce/National Bureau of Standards. This document can be obtained from the National Technical Information Service, Springfield, Va., and provides much more detailed information on the subject of AC power and grounding in ADP and similar applications.

The input frequency will be specified either as a range of values, or as a particular value with a tolerance range. Ferroresonant type of supplies are most often single frequency and might be specified as 60 ± 3 Hz or 57–63 Hz, for example. The same may be true for any type of power supply that uses a 60 Hz only AC fan for cooling. Linear power supplies may be specified for a wider range of frequencies, for example 47–63 Hertz, or 47–1000 Hz. Fans are available for 50–60 Hz operation, and are quite often used in switching type power supplies. Direct current fans may also be used, which draws a small amount of output power from the power supply. Because of small bias supply transformers, some switchers may be restricted to a narrower frequency range than one might expect.

High Line

The maximum steady state input voltage is determined by the voltages that are specified for the application, which are usually determined by the country in which the product will be used. In the United States, power companies typically maintain the voltage at 120 Vac + 5.8% / minus 13.3% which makes the high-line value 126.96 V in the United States.

Power Disturbances

Figure 1-2 shows the power disturbances described in the following paragraphs.

---

[1] See ANSI C*$.1 American National Standards Institute.

[2] See also ANSI X4.11.

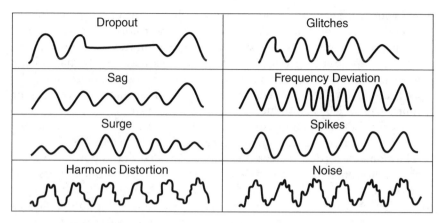

Figure 1-2.   Various Types of Power Disturbances

Surges are periods of overvoltage where the input voltage increases lasting more than 200 μs up to 16.7 ms.[3] Surges may occur when a very large load is suddenly removed from the line. This is especially true where the load is at a great distance from the last point of regulation. Surges can go higher than 140 V on a 120 V power line. Many power companies will make minor system changes if they are aware of controllable or alterable conditions that cause surges. Testing a power product's ability to withstand surges can be important for a number of reasons. Power in some rural areas or heavily industrialized areas may not be as well regulated where older equipment is in use. Power outside of the United States may be far less dependable and even erratic in some of the younger and underdeveloped nations on this planet, while most European nations have good quality power.

Dropouts or **Missing cycles** are just what the name implies, a period of missing input power. Testing for the capability to withstand missing cycles or input power dropouts of specified duration must be done under low-line conditions, unless otherwise specified. This test will prove the ability of a power supply to maintain outputs to specifications while undergoing a period of missing input power.

Spikes are transient periods of overvoltage during which the voltage increases for up to 200 μs.[4]

Bump (not shown) is a term used by many power company engineers to identify voltage increases of a few to several seconds. A Bump is a long surge. These are usually events that occur at a rate slower than the power

[3] EPRI JOURNAL, November 1985, p. 8.
[4] EPRI JOURNAL, November 1985, p. 8.

line frequency. They can be caused by: branches falling or being blown into power lines, car and truck accidents where a power pole is struck, squirrels climbing onto feeder lines, line switching, regulation transformers stepping switching taps, and other reasons. These bumps may also be the voltage surge caused by a power system recovering from a fault. Understanding the specifications for the intended application usually includes the country or countries in which the product is intended to operate.

Sags are similar to bumps, except that they decrease the value of the line voltage rather than increase it, and they last from 67 ms to 1 S.[5]

Harmonic distortion is created by any device that draws power from the line in a nonsinusoidal fashion. The current draw that is out of phase, or not continuously in phase with the voltage waveform contributes to the distortion of that voltage waveform. This distortion is due to the various inductive, capacitive, and resistive components of the various segments of the AC power distribution system and its component parts, especially including the loads placed upon it.

Glitches occur as single events or a small cluster of events. They may be caused by power fluctuations due to various loads being applied to the power system, or being removed from the system. These events are usually very fast in nature. A light bulb being turned on via a wall switch may cause a small glitch. A **spike** is a large glitch.

Noise is similar to glitches, except that they are less random, as they may be caused by electromechanical devices such as the brushes of a motor. Noise may also be created by fast switching transistors in power supplies.

Frequency deviation is a concern when the power source is not as well regulated as the normal AC power in the United States. This may be caused by power sources such as: emergency power sources, portable generators, private power systems, and power outside the United States and Canada, especially in less developed nations.

Testing can be accomplished by using transient and line noise signal sources that are available from several vendors that generate various line transients with varying degrees of control. Some testers and AC sources have the capability of providing missing cycles and other limited transient testing. Other more expensive systems can meet various agency testing standards. Be sure to study which standards your equipment will be required to meet prior to specifying test equipment.

Maximum Transient Input and Duration   Voltage transient tolerances are also usually specified. Typically transients will be specified with a peak voltage and a pulse width. Several manufacturers produce test equipment

[5] Ibid.

Table 1-2. A Comparison of Some Basic Circuit Protection Methods

Device	Speed	Sensitivity or Protective Level	Energy Handling Capability Alpha	Stability	Comments
Gas-filled surge arrestor	Very fast	Good	High	Excellent	Low voltage ionization levels, high discharge capability, excellent response time, good life
Metal oxide varistors	Very fast	Very good	High 25–70	Excellent	Excellent clamping, high discharge current level, very fast, low leakage
Air Gap	Fast	Poor	High	Poor	High peak power, small, low cost
Carbon gap	Fast	Poor	High	Poor	Noisy, requires maintenance, fairly fast, poor AC discharge
Selenium rectifiers	Fast	Good	Low 5–15	Poor	Very low energy rating
RC Networks	Medium fast	Very Good	Low about 1	Very good	Fast, low energy only
Zener Diode	Medium	Good	Low 35	Excellent	Low energy, good at low voltages, fairly fast
Circuit Breaker	Very slow	Good	High	Good	Very slow, big, expensive
Fuses	Very very slow	Good	High	Fair	Very slow, one-shot, even the fastest fuses are slow
Low Q Shield Beads	Very fast	Very Good	Low–Med.	Excellent	Low energy levels, very high speed, low cost

specifically for testing equipment tolerance to transients. Typically, transients of 20,000 V of very short duration will occur in an industrial area hundreds of times a day. In residential areas the same transients will be found, but there will be fewer of them.

Low Line

The specified low-line value has changed over the years as power consumption has increased. In the United States, power companies typically maintain the voltage at 120 Vac + 5.8% / minus 13.3% which makes the low line value 104.04 V. It is important to test power supply designs at minimum input voltage and at the lowest specified power input frequency at full load. This test presents the greatest stress to the filter capacitor at the input to the regulator. Power supply specifications often use a lower voltage to take foreign and brownout operation into account.

Survivability

Sometimes equipment will be specified for operation beyond specifications without damage. Equipment will not be expected to meet output specifications, but will be required to not suffer damage during this time. An example of this situation is operation during brownouts where specifications call for no damage when input power falls well below the minimum specified value for proper operation. Another example would be start-up after extended exposure to extreme cold temperatures.

In-Rush Current

Input current is determined to a large extent by the input circuitry that keeps the outputs turned off until the input is stabilized. Designers may use current limiting thermistors, SCRs, or relays to limit the inrush current. In each design, the type of input current limiting must be examined to determine to what extent inrush current is limited. The type of load connected to the power supply at turn-on may also affect some designs inrush current. For power supplies that operate at more than one frequency, the surge rating must consider all input frequencies. Typically, the inrush current will be tested at 230 V 50 Hz and 115 V 60 Hz. In most cases increasing the input power frequency will cause the inrush current to decrease, especially at frequencies several times the lowest input frequency. In measuring inrush current, the most critical factor is the impedance of the source of power to the unit under test, and the connections to it. Some power supply specifications may specify the minimum and maximum line impedance that it can operate from.

Power Factor

Power factor testing is becoming increasingly important due to the attention being paid to the harmonic content in power systems. This is largely due to the increasing consumption of power that is being drawn by loads, including power supplies, that do not draw current in a sinusoidal fashion. This trend, if it continues, may result in the regulation of the harmonic content of the current waveform drawn, that is, the power factor. There are several ways to measure power factor, depending upon the test equipment available. The power factor is the ratio of true to apparent power drawn by the product under test, expressed as a decimal. In many of today's power supplies, power factor is more complex than just a phase difference between voltage and current. Distorted current and voltage waveforms caused by nonsinusoidal current draw, create a power factor that cannot be easily measured with a simple oscilloscope. Some power supplies employ power factor correction circuitry to reduce the amount of distortion in the current waveform by forcing current waveforms to closely track the voltage waveform.

For more information on power factor correction (PFC), see the following application notes available from Micro Linear Corporation, 2092 Concourse Drive, San Jose, Ca. 95131. These notes may be requested by phone or Fax at: Phone: 408-433-5200, FAX: 408-432-0295 http://www.microlinear.com.:

- "Application Note 16   Theory and Application of the ML4821"
- "Application Note 33   ML4824 Combo Controller Applications"
- "Application Note 34   ML4824, A Novel Method for an off-line PFC-PWM Combo Controller"
- "Application Note 11   Power Factor Enhancement Circuit"

There are various methods used for three Phase power factor correction. One of the simplest and most cost effective is the one described in a paper that can be found in *IEEE Transactions on Power Electronics*. Vol. 6. No. 1, January 1991, titled "An Active Power Factor Correction Technique for Three-Phase Diode Rectifiers." The authors are *A. R. Prasad, Phoivos D. Ziogas,* and *Stefanos Manias.*

One additional benefit of power factor correction is that the regulating boost converter can be designed to alleviate the requirement for a voltage range switch.

One insurance company in a major metropolitan city, carefully planned its new computer installation, making sure that sufficient power was available for the computer room and the new computer system. Shortly after installation, when the substation transformer blew up, it was discovered that a thousand new computer terminals in the building had not been taken into the equation. Each terminal with 1 A of RMS current, but an actual 10 A peak value every half-cycle, equates to 10,000 A peaks every half-cycle for a 1000 A draw. And that's just for 1000 terminals. After you add in all the other office equipment, fluorescent lighting, and industrial loads, the once sinusoidal waveform can become quite distorted and "dirty." The harmonic comment is high and manifests itself in the transformer in the substation.

Line Regulation

Line regulation is the ability of the power supply to compensate for changes in the line voltage while maintaining the output voltage within specifications. Line regulation typically is measured from 100 or 105 V to 125 or 130 Vac. Additionally you will find brownout specifications down to 70 or 80 V region.

Warmup Time

Once power is applied and the power supply is performing, a period of time may be specified to allow the power supply to meet all specifications. The starting temperature for this test is often the specified storage temperature limit for the power supply, from which the unit may be powered up, without damage, but may not necessarily meet all specifications for output voltage, current, or other specified parameters. Testing may be accomplished by measuring defined parameters to specification after a given period of time.

Start-Up Time

Start-up time is the time from the application of input power until all outputs have stabilized at the specified values. The assumption is made that the unit under test is at specified operation environmental conditions.

Hold-Up Time

Hold-up time is generally specified for power supplies where there is a requirement for power to remain available to the load after the loss of input

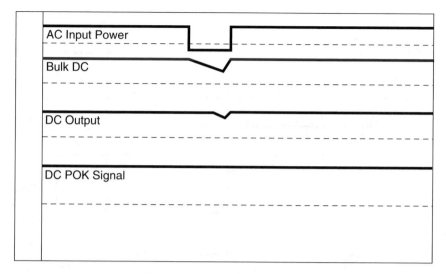

Figure 1-3. Hold-Up Time

power. The hold-up time is defined as the time between loss of input power and the time at which an output falls out of specification (Figure 1-3).

In Figure 1-3, T0 is defined as the start of power failure. T1 is defined as the point in time when the AC OK Signal is lowered. T2 is defined as the point in time when the DC POK signal changes state due to the DC output going out of specification. Holdup time is specified as the time period between T0 and T2. The time from T0 to T1 is the time required by the power fail circuitry to determine that the power supply has actually had an input power failure. The power fail signal is sometimes called AC-OK, Power GOOD, or AC-GOOD. The time between the detection of an input failure and the point where the output voltage goes out of tolerance is the time required by the powered device to prepare for shutdown. The DC-OK, AC-OK, and POK signals may be true in either state as specified.

Dropout/Withstand Time

Dropout time is specified as the longest duration of power off that a unit can withstand, and still hold the outputs within specification. When a power supply has an AC-OK circuit, the withstand time is generally the time from the NOT-AC-OK until power falls out of specification at full load. In Figure 1-4 the Bulk DC decreases during the dropout, but the output remains in regulation to specification. Any variation in the output voltage will be within the specified output tolerance.

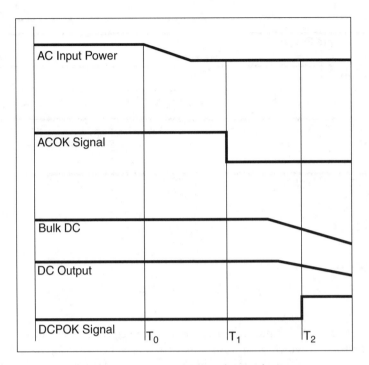

Figure 1-4. Dropout/Withstand Time

Leakage Current

Leakage current is the current that flows into the power supply from the primary input to either ground or the outputs. In some applications, for example, patient-connected medical equipment the leakage current must be very low to prevent the death or injury of a person connected to it. The legal aspects of testing power supplies used in medical applications[6] also needs to be examined. Perhaps the most stringent measurements of leakage currents are the demanding requirements spelled out in UL-544 for patient-connected equipment. Regulations covering other types of equipment may require leakage measurements as well.

Consumer goods are also subject to various regulations, depending upon the device and its application. Some of the important Agency Regulations are: Underwriters Laboratories (UL) UL-544, UL-114, UL-478, Canadian Standard Association (CSA) CSA 22.2 Nos. 143 and 154, International

[6] See "Testing Instrumentation for Medical Equipment," a three part series in the *Journal of Clinical Engineering*, Vol. 2, No. 3, July–September 1977, p. 242.

Electrotechnical Commission (IEC) IEC 380 Class I and II, IEC 435 Class I and II , Verband Deutscher Elektrotechniker (VDE) VDE 804 Class I and II, and VDE 806 Class I and II.

HIPOT Testing

The term HIPOT testing is a throwback from the early days of motor generator testing and refers to the high potential voltage applied between windings and the case or ground. This potential was applied to find any leakage paths or points where breakdown would occur.

Some modern HIPOT testers of today are designed such that they can destroy the circuits under test. Take care in determining that your test equipment is appropriate for your usage.

WARNING

When using HIPOT testers be aware that the type used for electrical apparatus such as motors and generators have been known to damage electronic equipment under test. This is due to the design including an inductor or radio frequency (RF) choke, with a neon lamp across it to indicate that a breakdown has occurred. This couples a very narrow but very-high-voltage pulse into the equipment under test. Sensitive circuits will be damaged or destroyed by this pulse under most conditions.

HIPOT Testers

The HIPOT tester by Hipotronics,[7] or its equivalent is the instrument of choice for electronic equipment, or equipment containing electronic circuits and components. This design ramps up a voltage to a preset level, and if excessive leakage is noted, the tester automatically crowbars the source voltage. Equipment specifications usually spell out HIPOT testing, or refer to an applicable standard that specifies it.

Grounding Schemes

There are various grounding schemes used in electrical equipment. The three that we will consider briefly here are: grounded, isolated, and low leakage.

---

[7] Hipotronics, Inc., Route 22, PO Drawer A, Brewster, NY 10509 phone: 914-279-8091.

- **Grounded** In the most common grounding architecture in America, the chassis is grounded along with any transformer cores and shields. Input RFI filter capacitors are also usually connected to this ground. As a result, some ground current will always flow. Component selection choices will determine how much ground current will flow in your product. The purpose of this grounding is to drain off any transient or lightning currents to ground, thereby protecting equipment.

- **Low leakage** In this scheme, used extensively in medical equipment, leakage current is intentionally kept low. Refer to the regulations applicable to that equipment for testing requirements.

- **Isolated** In some systems, especially smaller low power units, full isolation is achieved without any ground at all. Filtering is connected across the line, with the design ensuring low coupling between the AC line and the loads.

- **Foreign** Many foreign countries require testing to their specifications. Many European countries using 220 Vac power have leakage detectors (Ground Fault Detectors) built into their electrical systems. Most European systems do not use ground as we do in the U.S., but float both sides of the line with ground fault detectors in place. Any current flowing to ground will trip the circuit. In some cases this precludes the use of large filter capacitors between the line and ground to solve radio frequency interference (RFI) problems. Testing of a product to meet many of these different specifications is typical. It is common to find these systems without the typical ground conductor found in most U.S. products.

Ground Current

Some equipment, especially medical equipment, requires a low level of ground current. Depending upon the equipment's application, maximum ground current could be as low as 100μA or lower. The applicable UL or other specifications will usually define the method of testing.

Common Grounds

Common grounds are parts of a circuit or circuits in which grounds are connected together at some point. This does not mean that they are interchangeable, since each may carry currents of different types and levels. Some common ground points may have a few ohms of resistance intentionally designed into the circuit, to force the majority of the current to flow in the desired paths.

Ground Loops

A ground loop is a circuit where a circuit is connected via two or more different paths to ground. The resulting parallel circuit can cause an unwanted current to flow in a ground path resulting in a voltage being developed across the resistive and reactive elements of the path. The voltage developed, even a few microvolts, can sometimes cause feedback and oscillation, rendering the circuit or system either dysfunctional or nonfunctional.

Isolated Grounds

Some power supplies or subsystems may have isolated grounds where the various grounds are isolated and only tied together at a single point. Some of these grounds may not even be tied directly together but may remain at some other voltage level.

Ground Testing

In some applications testing of ground connections is required. Some large cities have regulations requiring testing beyond Underwriters' Laboratories (UL) and other standards. Such regulations will have to be dealt with as they are encountered.

Other Specifications

Custom power supply specifications will often contain peculiar items that will need special attention in testing. This is one of the reasons for the custom power supply marketplace, which gives users the ability to obtain power products designed for special applications.

DC Input

Some power supplies have DC inputs either for battery backup, or as the main input as in DC to DC converters. The following are specifications where the DC input is different from the AC input specification.

HI Line

Hiline is defined as the maximum input DC voltage that can be input to the power supply. This voltage includes all variations of the input voltage, along with ripple. It does not include voltage transients and noise.

LO Line

LO Line is the minimum input DC voltage at which the power supply will continue to meet all output specifications. This may or may not include the ripple voltage.

Start-Up Time

Start-up time is the time from the application of DC input power until all outputs have stabilized at the specified values. The source impedance could become an important value if the start up time specified is very short. The assumption is made that the unit under test is at specified operation environmental conditions.

Tolerance to Ripple

Power supplies that operate from a DC line may be exposed to input power which contains large amounts of ripple. Supplies operated from an AC line may be exposed to poorly regulated power which varies in voltage. The size of the bulk input storage capacitor (if used) will aid in the ability of the power supplies to tolerate ripple.

Tolerance to Noise

Tolerance to noise is determined by design of the input filter, physical shielding, physical layout, component placement, and many other factors. The size of the input filter (if used) will aid in the ability of a power supply to tolerate EMI, RFI, and noise from external sources. This same filter will also reduce the amount of EMI, RFI, and noise that the power supply will radiate into the line.

Output Specifications

Power supply output specifications are generally the most critical area of concern. Output tests are performed on the various outputs of power supplies. Most often the outputs are DC, and these power supplies are sometimes referred to as AC to DC converters. Some power supplies have DC inputs and AC outputs and are usually referred to as inverters. Other outputs may include bias voltages, clock signals, AC and/or DC low, AC and/or DC hi signals, Power OK signals, sync signals, etc.

DC Outputs

Static Load Regulation

Static load regulation is the regulation of outputs under static or unchanging conditions. The outputs are tested to determine that the power supply can maintain the specified output voltages within the specified tolerances, under the range of load conditions specified for the product. If an output is specified as 5 V ± 2% from 1 to 20 A, then the output must be tested through the range of 1-20 A. Engineering would probably test at many more points than would be used for a test in the manufacturing environment. Manufacturing would certainly test the maximum value, and usually the minimum value, and perhaps even some nominal value, but hardly ever more than that unless such testing is specified for a particular reason. In each case, the current is set to a specific value, the output is given some number of milliseconds to stabilize, and then the output voltage is measured.

Dynamic Load Regulation

Dynamic load regulation is the ability of a power supply to regulate the output voltage under dynamic conditions. Usually this implies that the load is not steady state, but rather is changing between two values at some specified rate. This specification is closely linked to the AC response of the regulation control circuitry. One area where dynamic loading is often specified is for applications where the load is a motor in a disk drive. It is not uncommon for a power supply specification to contain a description of the current waveform. In some computer and industrial controller applications, the load current may vary in discrete steps. Whatever the load, custom power supplies are often designed for those special load conditions. Testing to those load conditions often becomes a real requirement.

Some loads are designed to provide those custom load requirements during testing. One company [3H Industries, which was purchased by Analogic, the product was subsequently dropped] developed loads called the Swinger 1 and Swinger II which had the capability for setting the load current with minimum and maximum values as well as independent rise and fall times. The maximum rate of change was 20 A/sec.

Short-Term Drift (Warmup)

Warmup drift is a measure of the output change due to initial warming of the power supply over some period of time. Typically, this parameter is measured by turning on the power supply with full rated load and measuring the output after it is stabilized (10–20s). After some predetermined time, usually 15–60 min., a second reading is taken. Warm-up drift is then calculated as follows:

$$D = \frac{Vta - Vtw}{Vtw} * 100\%$$

Where:

  Vta = voltage at ambient temperature
  Vtw = voltage after warm-up
  D = drift (in % of output)

Some causes of short term drift are:

- Thermal excursions in the environment
- Inadequate circuit design
- Component(s) that fails to meet specification
- Inadequate physical layout and/or package design
- Large changes in load current, beyond specified limits

Long-Term Drift (Voltage)

The major difference between short- and long-term drift the time that the test is run for. While only minutes of time are discussed in short-term drift, the time for long-term drift is usually 1 year or product lifetime, depending upon the specifications. Testing this parameter will require a measurement device that itself has excellent long-term drift specifications.

Some causes of long term drift are:

- Severe thermal excursions in the environment
- Component aging
- Inadequate circuit design
- Component(s) that fail to meet specification
- Inadequate thermal design
- Inadequate physical layout and/or packaging

Thermal Drift (Voltage)

Power supply outputs will change due to a change in ambient temperature. Usually this is not a significant problem, however in some applications,

Thermal drift is a critical specification. Testing can be done locally or may be performed by an independent service. Attention must be paid to the details.

Output Impedance

Many specifications for power supplies ignore the impedance of the outputs. The output impedance of a power supply is important information when choosing standard off-the-shelf power supplies for a particular application. Some manufacturers are starting to provide this important information in their specifications.

Overvoltage

A common option for a power supply output is overvoltage protection. This option can be implemented in many ways, but the two most popular are the crowbar circuit and the shutdown circuit.

Crowbar overvoltage protection occurs when a short is placed across the output to prevent the load from suffering from the affects of an overvoltage condition. This is often combined with a shutdown of the power supply either directly or as a result of the short of the crowbar. The crowbar was most popular in series linear regulators where a shorted pass transistor would couple the full input voltage to the output. In logic circuitry the results were disastrous.

Shutdown overvoltage protection occurs when the power supply shuts itself off after sensing an overvoltage condition. This method of protection is most often used in switching regulators where there is transformer isolation which prevents any accidental overvoltage condition on the output due to component failure.

Cross-Regulation

Cross-regulation is the specified regulation of an output in a static condition, while a different output load is varied to a specified range. While the goal is that all outputs are completely isolated from each other, in reality each output is affected by the others. This specification determines to what extent this situation is allowed, and the purpose of testing is to show that the specification has been met.

Isolation

Many power supplies are used for the isolation of the loads from each other and from variations of any type in the input power. The isolation can be specified as AC, DC, and/or noise. The noise isolation is often specified in terms of capacitive coupling between outputs.

- DC isolation is often tested via HIPOT testing or a specified resistance value measured in megohms, at a specified voltage value such as 500 Vdc.
- AC isolation is often tested in the same way that DC isolation is tested, via HIPOT testing or a specified resistance value measured in megohms, at a specified voltage value such as 500 V.
- Noise isolation is usually achieved by additional shielding designed into the power transformer and is quite often specified with a capacitive value in picofarads or isolation in decibels.

EMI / RFI / ESD

The standards that power products must meet, either on their own or as a part of an overall system, can be myriad. The goal is twofold, to prevent the power supply from emitting undesirable signals by radiation or conduction, and to prevent external electromagnetic or RF signals from interfering with the operation of the power supply or system, or some other system nearby. Electrostatic discharge (ESD) has become a concern for semiconductor products using field effect transistors in almost any form. This includes MOSFETs, CMOS, and many controller integrated circuits (IC).

There are many agencies whose regulations, and standards may be specified for a particular product, and against which the product will be tested. These agencies include: VDE, UL, CSA, Federal Communications Commission (FCC), and International Electrotechnical Commission (IEC), among others.

Conducted Signals

Conducted signals may enter or exit the power supply through the power input leads, the power output terminals, through a faulty or improperly connected ground, or even through the case of the power supply. Testing is most often performed by certified testing houses, or by the agencies themselves.

Radiated Signals

Radiated signals may enter or exit the power supply through a faulty or improperly connected ground, through the case of the power supply, or through improperly connected or routed wiring in open frame power supplies. Testing is most often performed by certified testing houses, or by the agencies themselves.

Electrostatic Discharge (ESD)

ESD may enter the power supply through the power input leads, the power output terminals, through a faulty or improperly connected ground, through knobs and hardware, or even through the case of the power supply. During manufacturing, the components and sub-assemblies are even more susceptible to ESD. In qualifying a vendor to manufacture power products for you, it is often the norm to qualify the process used, including ESD protection and training at manufacturing sites. Testing is most often performed by certified testing houses, or by the agencies themselves.

Sequencing

Often timing constraints that are placed in a specification demand that certain outputs will come up in a specified order. The same may be true for the removal of power as well. More recently designed products and systems seem to have gotten beyond this requirement, largely due to better designed memory chips and systems, but some applications may require it.

Start-Up and Shut-Down

Startup of systems where a certain power comes up before another is common. Sometimes a tracking requirement is placed in the specification so that one output tracks its rise and fall with another, or before or after another. It may even be specified that such sequencing takes place during overcurrent and shorted conditions.

Current Limiting

There are several forms of current limiting. We will examine the major ones here.

Straight Current Limiting

Straight current limiting limits the current to a maximum value. As shown in Figure 1-5, when the current reaches the maximum value the current is

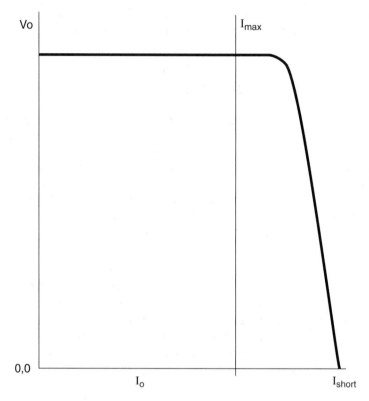

Figure 1-5.   Straight Current Limiting

limited as the voltage falls to zero. Notice that the output current is still rising all the way to the shorted condition. Depending upon the design method used, the slope of the output voltage will vary. This method is not often used in high-current linear power supplies due to the power that is dissipated in the power supply during current limiting operations and due to the power that could be dissipated in a partial shorted condition.

Foldback Current Limiting

Foldback current limiting is usually the best choice over standard current limiting techniques, particularly where output device power dissipation is a concern. The purpose of foldback current limiting is to cause the output current to decrease whenever the power supply has first exceeded its maximum value and when the load resistance is lower than specified for maximum

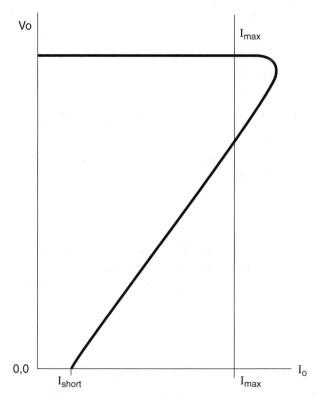

Figure 1-6. Foldback Current Limiting

load. In Figure 1-6, the output voltage is available up to the maximum current Imax.

The output current then folds back with decreasing load resistance to a value of I short where the load resistance is essentially zero. Some designs will latch in this condition and require some type of reset, but most will automatically reset themselves upon removal of the short. In practical applications, regulation may be seen to degrade as Imax is approached. Some designs will exhibit regulation degradation at 80% of Imax while others will regulate even closer to the Imax value.

Latching Current Limit

Some designs will latch in the overcurrent condition and may require some type of action to reset them. Most power supply designs will automatically reset themselves upon removal of the short. Latching should not be assumed unless specified.

Impedance Current Limiting

In some designs, both regulated and unregulated, overcurrent protection is achieved by careful design of the power supply so that the impedance of the power supply output is such that excessive current cannot be drawn. These designs may even allow long term shorted conditions to exist without any damage to the power supply.

No Current Limiting

For some applications, the cost of current limiting either cannot be justified, or the function of current limiting is provided by the load circuitry or fuses. Care must be exercised in testing these units so as not to unintentionally destructively test them beyond their specified limits.

Efficiency

Efficiency is a measure of the power consumed in the process of power conversion. For DC-DC converters, it is simply the Watts Out divided by the Watts In, and multiplied by 100%. For AC-DC and AC-AC converters, the measurements process requires that careful attention be paid to the RMS measurement of both voltage and current, as well as the phase relationship between them. For many switching power supplies, a good indication of losses can be seen by shorting the output(s) and measuring the input power.

Ancillary Outputs

Many power supplies have ancillary outputs for many reasons. The following are some examples:

- AC OK
- DC OK
- Power OK (POK)
- Real-Time Line Clock (RTLC)
- Other signals

AC OK

AC OK is a signal that gives the status of the AC power input to the power supply. Often this signal is connected to a field effect transistor (FET) that will conduct without biasing, so that the status of a power supply with no power can be determined. The specifications will usually give the timing relationships that are required.

DC OK

The DC OK signal is just that, a signal that shows the status of all (or in some cases, specified) DC outputs to be within specification for the required timing which must be tested. In some computer systems the AC OK and DC OK signals tell the computer that power is available to continue operation, but to prepare for power loss.

Power OK (POK)

When a POK or Power OK signal is used it may be a combination of both the AC and DC Ok signals. Be sure to check the specification for the required action and timing that must be tested to verify the design to specifications.

Real-Time Line Clock (RTLC)

A RTLC signal is synchronized to the AC line frequency to provide a timing standard to other circuitry in the system. There are various types of circuits employed for this task, depending upon the accuracy that is required for the application. Some of the better designs may include Tee notch filters with a Q low enough, or sufficient bandwidth, to span the frequency range specified for the power supply AC power input. This type of design is generally more accurate as it ignores noise spikes.

NOTE

Remember, Q is X_l/R.

Other Signals

Carefully check the specifications for any other special signals which may have been specified. Other signals may include System Clock signals in switchers, reference voltages, crowbar activation signals, and remote shutdown sensing.

Ancillary Inputs

Many power supplies have ancillary inputs for any number of reasons. They may include the following:

- Remote control
- Remote sensing
- Test
- Reset signals

Each of these specifications must be examined carefully to determine the type of testing to be performed at engineering, manufacturing, and field service locations.

Military Specifications

It is especially important to get a copy of every specification referenced in any military power supply specification. It is normal to then obtain those specifications referred to in the first set of specifications. This process must be repeated until all the necessary specifications are in hand. Those who work with the military and military contractors on a regular basis will probably have most of the required specifications on file. Give these specifications the full attention that they deserve.

Quality

Quality programs such as the Motorola Six Sigma program are becoming increasingly popular. These programs can have considerable effects on manufacturing and engineering processes. Check specifications carefully for reference to any of these programs. Don't underestimate them, but fully research their potential impact on your company and you will be a winner.

Burn-In/Environmentally Controlled Stress Test

Some power supply specifications detail the required level of burn-in or environmentally controlled stress testing (ECST) that is to be done on both the pilot run and production versions of the product. when specifications are unclear and generic, it will pay to request an addendum to the specification nailing down these points. Specifying a process that fits neatly into your existing manufacturing process may be possible, but be prepared to expand your process, if necessary, to meet the customer's needs and desires. Many specifiers of power supplies do not really understand what burn-in is and how it should be used. Be sure to read Chapters 4, 5, 6, 10, and especially Chapter 16 where burn-in and environmentally controlled stress testing are discussed.

Life

Life Specifications

Occasionally a power supply specification will include life requirements, usually stated in Mean Time Between Failure (MTBF), stated in thousands of hours. These ratings can be obtained by several methods, including calculated, accelerated testing, and actual data.

Calculated

Using Military Standard 217b or some other method of calculating the MTBF of a product will produce an estimated value of the actual MTBF figure.

Accelerated Testing

Accelerated life testing is often performed on products to give an indication of the expected lifetime of the product. A small sample of product is tested at conditions above the normal ambient conditions to demonstrate the calculated or expected MTBF figures. Testing laboratories are often used for this task.

Historical Data

Feedback from failed units in the field, if available, can be a good indication of MTBF. The problem is that it takes years to accumulate this type of data. For some designs that are in service for an extended period of time, this may be an option to update and prove the actual MTBF figures of the product. Some power supply specifications may in fact spell out how this is to be accomplished.

AC Outputs

Alternating current power sources or power supplies that have one or more DC outputs may have additional specifications for AC outputs. Here we examine the output specifications that may be different from the DC output specifications.

Output Impedance

Output impedance is a complex measurement that is seldom specified for AC power supplies. What is normally specified is the maximum power output in watts with a derating curve or formula for power factors other than 1, at various output voltages. Typical ranges may be 100% of the rated volt amperes (VA) output power at a power factor of 0.30 and full output voltage, an output of 80% of the rated VA output power at a power factor of 0.05 and full output voltage, or an output of 150% of the rated VA output power at a power factor of 1.0 and full output voltage.

Peak Repetitive Output Current

For AC power supplies the output current when the load is nonresistive or has a power factor of less than unity. When the load of the AC power source in another power supply where the peak repetitive current can be many times larger than the average current, this specification is critical in understanding the expected response to the load presented to it.

Power Output

The power output is most often specified for the various output ranges and load power factors that the unit is expected to see at the load terminals.

Frequency

The frequency is the range of operation or the specific fixed frequency of the AC output at full rated output power.

Total Harmonic Distortion

The total harmonic distortion is often specified for the various load conditions of various power factors.

Response Time

Alternating current outputs on power supplies require a certain amount of time to respond to changes in load levels. This specification provides the value of that time. If the output voltage is also variable or controllable, this specification may also refer to the time required for a change in the control signal level or state to be realized in the output.

AC Noise Level

The AC noise level is a measurement of the output signal that is not harmonically related to the fundamental output frequency sinusoidal waveform.

Overload and Short Circuit Protection

Overload conditions are usually specified in at least two values. Instantaneous or peak and average. The instantaneous value is specified as a higher value to allow the output to provide power during startup of motors and other reactive loads. When the load of the AC power source exhibits a peak repetitive current many times larger than the average current this specification is critical.

Frequency Stability

For some applications, the stability of the output frequency may be critical.

Frequency Stability versus Time

Testing the frequency stability to the specification versus time is often required. Due to the lack of standards, the criteria of time must be included in the specification.

Short-Term Stability

Typically the short-term stability is a measure of the immediate variations in the output frequency. Where possible, the method of testing should be defined in the specification.

Long-Term Stability

Long term stability or drift is a measure, over an extended period of time, of the output frequency.

Frequency Stability versus Temperature Changes

The frequency stability versus temperature changes specification defines the output frequency change due to specified changes in the ambient temperature. Due to the thermal mass of some parts of the product, time for thermal equilibrium in some components may be required. To be accurate and repeatable, the specification must define the thermal excursions and the time required before measurement is made. This is another area where specification of the test method and procedure in product specification can really prevent future problems.

Frequency Stability versus Input Voltage Changes

The frequency stability versus input voltage changes specification defines the output frequency change due to specified changes in the input voltage. This quality is usually specified at full power and at low-line, nominal line, and high-line input voltages. Testing is usually performed the same way.

Frequency Stability versus Load Current Changes

The frequency stability versus load current changes specification defines the output frequency change due to specified changes in the load current as well as the power factor of the load. Specific load currents at specific power factors are normally included in the specification.

Waveform Distortion

This waveform specification defines the output waveform distortion due to specified changes in the line voltage and load current.

Spectral Purity

Spectral purity is a specification that defines the purity of the output frequency and is most often specified at minimum and maximum load conditions.

EMI/RFI Specifications

Power supplies of any type are generally required to meet certain standards for radiation of interference via either conduction or radiation of unwanted signals or both. In the United States, the Federal Communications Commission (FCC) regulations are the standard, however; products that may be destined for foreign markets will need to meet the applicable foreign standards unless that country relies upon the product meeting the FCC or other specified standard. During testing, both conducted emissions and radiated emissions will be measured.

Annotated Bibliography

1. Allen and Segall, *Monitoring of Computer Installations for Power Line Disturbances*. Institute of Electrical and Electronic Engineers (IEEE) Conference Paper C74 199-6, IBM Corporation. An informative summary of the magnitudes and frequencies of occurrence on normal mode power source disturbances and outages observed at numerous ADP sites.

2. ANSI C2, American National Standards Institute, *National Electrical Safety Code*, IEEE This book covers safety and grounding issues related to the generation, transmission and distribution of electrical power. Do not confuse this with the *National Electrical Code*, NFPA-70, for premises wiring (see below).

3. ANSI X4.11, American National Standards Institute, *Operating Supply Voltage and Frequency for Office Machines*. Lists the frequency range and voltage ranges for which office machines are to be designed.

4. Shepard, Jeffrey D., *Power Supplies*. For a good overview of government and industry standards, see Chapter 8 of this book.

5. ANSI C84.1, American National Standards Institute, *Voltage Ratings for Power Systems and Equipment (60 Hz)*. Lists nominal voltage ratings together with operating tolerances for electrical supply and utilization systems. Also includes data on principal transformer connections and grounding.

6. Bodle, Ghazi, Syed, and Woodside, *Characterization of the Electrical Environment*, University of Toronto Press, 1976. A handbook of information on effects and characteristics of overvoltage surges, lightning, power interference, corrosion, electromagnetic pulses, earth potential gradients, and electric shock are discussed.

7. FIPS Pub. 94, *Guideline on Electrical Power for ADP Installations*, September 21, 1983, Federal Information Processing Standards Publication, U.S. Department of Commerce/National Bureau of Standards.

8. Goldstein and Speranza, *The Quality of U.S. Commercial AC Power*, Bell Laboratories, Whippany, NJ, IEEE International Communications Energy Conference Proceedings, October 3-6, 1982. Statistical results of 270 months of data gathering at 24 ADP sites arranged in tables to show the number of disturbances of various types predicted per year, and the improvements expected from several types of power conditioning equipment.

9. IEEE Standard 142-1982, *Recommended Practice for Grounding of Industrial and Commercial Power Systems*, (the Green Book).

10. IEEE Standard 241-1983, *Recommended Practice for Electric Power Systems Commercial Buildings*, (The Gray Book).

11. IEEE Standard 446-1980, *Recommended Practice for Emergency and Standby Power Systems for Industrial and Commercial Applications*, (the Orange Book).

12. IEEE Standard 518-1982, *Guide for the Installation Electrical Equipment to Minimize Electrical Noise Inputs to Controllers from External Sources*. This guide contains descriptions of electrical noise sources and techniques for minimizing them.

13. NFPA 70 (ANSI/NFPA 70), *National Electrical Code*, National Fire Protection Association, Boston, MA.

14. Fisher, Franklin A.; Plumer, J. Anderson, *Lightning Protection of Aircraft*, National Aeronautics and Space Administration (NASA) Reference Publication 1008, Superintendent of Documents, U.S. Government Printing Office, Washington, DC, October 1977. This practical designers' handbook has theoretical and practical information and data on tests on preventing lightning currents from destroying and interfereing with the functioning of electronic circuits.

15. Teets, Rex M., *AC Power Handbook of Problems and Solutions*, Gould, Inc. 2727 Kurtz Street, San Diego, CA 92110, 1981 (3rd Ed., $4.00). A small handbook of nearly 100 pages describing computer power problems and solutions using power conditioning.

16. ———, *Transmission and Distribution Reference Book*, East Pittsburgh, PA, Westinghouse Electric Corp.

17. ———, *The Effects of Electrical Power Variations Upon Computers: An Overview*, U.S. Department of Commerce Domestic and International Business Administration, Stock No. G-0325-00025.. Superintendent of Documents, U.S. Government Printing Office, Washington, DC, July 24, 1974.

18. Underwriters Laboratories 478, *Standard for Electronic Data Processing Units and Systems*, Contains requirements and test specifications for ADP systems and subsystems.

2

Simulation

There are many different simulation tools available on the marketplace today. While some are based on proprietary schemes and algorithms and many based on Simulation Program with Integrated Circuit Emphasis (SPICE) or versions of it. One vendor's version has achieved the greatest results into the power supply domain by providing models of real parts including semiconductors, IC's, magnetic cores and more. While this chapter has been written with this tool in mind, it applies to most all spice-based simulations, and perhaps a few others.

While this is not an endorsement of any tool, the Analog Workbench is a tool that has apparently come the closest to meeting the needs of power supply designers. The Analog Workbench is perhaps the most complete of the high-end systems. The Spice-based Analog Workbench was originally designed by Analog Design Tools, who later was acquired by Valid Logic Systems. Valid Logic Systems enhanced and updated the tool before it was acquired by Cadence Design Systems. Cadence is located in San Jose, California and can be reached at 1-408-944-7690. Cadence has further improved the product to the present day version of the Analog Work Bench. More revisions are in the works in future releases, and Cadence is continually updating the model libraries. They have added more capability to the system including a IBIS model translator that allows the user to use IBIS models in SPICE. Perhaps the greatest addition to the workbench has been software that allows the inputting of desired results from a simulation, and then the software tests the circuit until it finds the optimal set of component values and parameters that will deliver the desired output. This can greatly reduce the debug time in the lab on new products! Less costly tools are available but with lesser degrees of accuracy.

Simulation in the Development Process

There is no one way to design, develop and manufacture power supplies, but there are several steps involved which are very common. These include specification, topology selection, circuit design, physical layout, test development, and others. Adding simulation into a development process can have great benefits, but necessarily brings with it new activities which must be planned for if these benefits are to be realized.

Suppose the original development process looked something like Figure 2-1. In this case, the bulk of the design work is done using breadboards and lab equipment. The engineer relies on his or her judgment and intuition to solve problems and to verify that these problems stay solved throughout the development process.

These fundamental activities do not automatically disappear when computer simulation is introduced. Instead, a few new activities appear as in Figure 2-2, and several existing ones become considerably expanded. The first new step, **modeling**, is to acquire simulation models for any elements of the design for which models are not already available. This step is important, for the rest of the computer analyses will depend on it. In some cases, "training the engineer to use the computer" may really be the first step, depending upon the engineer and the tools selected.

Simulation then plays an active role in the circuit design phase. Typically, much of the "what if" kinds of analysis done in design will be done in simulation and considerably less using breadboards. This means that few breadboards will be built and the role of a technician will be smaller. Under many circumstances, and especially in a power supply design, breadboards will not be eliminated completely and will remain an important part of the development process.

Much of the design analysis phase will relate not only to satisfying the product specifications, but also to ensuring that the product can be manufactured economically, and that it will not fail in the field after it is delivered to the customer. Computer analysis can play a major role here, and also can be used to assist in the development of product testing. More on this later.

A relatively recent innovation for power design is the ability to calculate and analyze the electrical impact of the ways the printed circuit board is physically laid out. Parasitic components of inductance, resistance, capacitance can be extracted from the board layout and added back to the circuit when desired. Thermal effects can also be modeled for components that are near heat sources, or that generate heat themselves. This is of particular

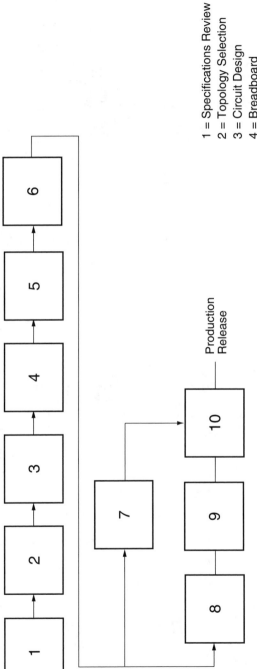

1 = Specifications Review
2 = Topology Selection
3 = Circuit Design
4 = Breadboard
    Low Power Sections
5 = Full Breadboard
    (n passes)
6 = Board & Product
    Layout
7 = Test Development
8 = Prototype
9 = DEBUG & Changes
10 = Pilot Production

Production
Release

Figure 2-1. Original Development Process

39

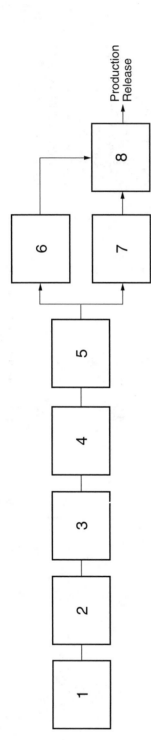

Figure 2-2. Modified Development Process

1 = Specification Review
2 = Topology Selection
3 = Circuit Design
4 = Device Modeling
5 = Simulation
6 = Test Developement
7 = First Pilot (Proto Type)
8 = Pilot Run

relevance since the major alternative to a computer analysis is to do an actual prototype board. For a design sensitive to layout effects, which may require several iterations, this can be extremely costly booth in real cost and in time to market impact.

Simulation Tools and Instruments

SPICE was developed at the University of California at Berkeley in the 1970's by Lawrence Nagel and Donald Pederson. Over the years SPICE has been extended, enhanced, and even re-architected both at Berkeley and by commercial software vendors to the point where today's state of the art tools bear little apparent resemblance to the early SPICE programs! Nonetheless, the basic SPICE "engine" algorithms, which involve direct methods solution of nodal admittance matrices, variable timesteps, etc., have proven again and again to be the most reliable, widely useful methods available even today.

There are three principal simulation modes common to all SPICE implementations, along with several specialized modes which vary from product to product. The three major simulation domains are **DC**, **AC**, and **Transient** analyses. In a DC simulation, the simulator finds the quiescent operating point of a circuit. An AC simulation calculates the linear frequency response of a circuit. Finally, a transient analysis simulates the activity of a circuit over time. Transient, or **time-domain**, is generally the mode in which engineers spend most of their simulation time, in part, because DC and AC simulation usually takes very little computing time, and in part because the nonlinear characteristics and time response of a circuit are often those that warrant the most investigation.

Other less commonly used simulation modes include noise simulation, distortion simulation, nonlinear AC simulation, and pole-zero simulation. Availability and functionality of these modes varies from vendor to vendor.

The Analog Workbench, from Cadence Design Systems, is a complete computer aided engineering system which includes an engine, SPICE PLUS, a schematic capture system, a graphical user interface, a large number of post processing and design analysis tools, and integrated PC board layout system, and a large library of models for use in the system. The user interface of the analog Workbench is mouse and menu driven and was designed to be reminiscent of familiar laboratory test equipment, or a "real" workbench. This was done to make computer analysis accessible to users who might be good analog design engineers, but who may not necessarily be simulation or UNIX experts.

Some of the simulated instruments associated with the Analog Workbench are:

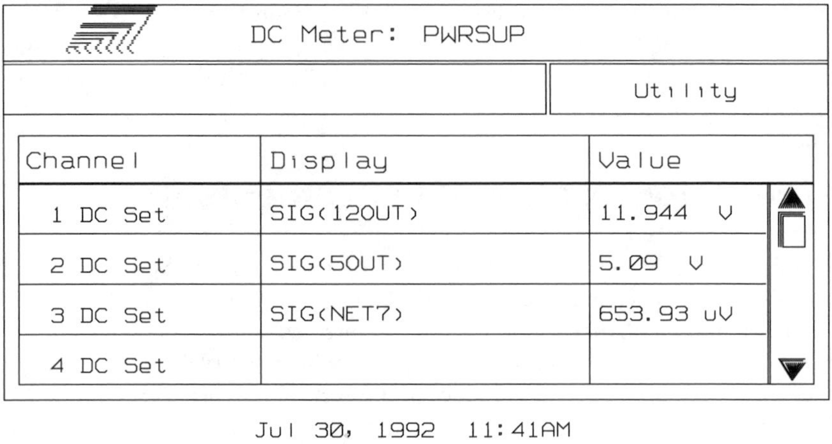

```
┌─────────────────────────────────────────────────────────┐
│  ▄▄▄▄▄                                                   │
│  ▄▄▄▄▄//      DC Meter:  PWRSUP                           │
├──────────────────────────────────────────┬──────────────┤
│                                           │   Utility    │
├────────────────┬───────────────┬─────────┴──────────────┤
│ Channel        │ Display       │ Value                   │
├────────────────┼───────────────┼──────────────────┬─────┤
│  1 DC Set      │ SIG(120UT)    │ 11.944   V        │ ▲▲  │
│                │               │                   │ □   │
├────────────────┼───────────────┼───────────────────┤     │
│  2 DC Set      │ SIG(5OUT)     │ 5.09   V          │     │
├────────────────┼───────────────┼───────────────────┤     │
│  3 DC Set      │ SIG(NET7)     │ 653.93 uV         │     │
├────────────────┼───────────────┼───────────────────┤     │
│  4 DC Set      │               │                   │ ▼▼  │
└────────────────┴───────────────┴───────────────────┴─────┘
```

Jul 30, 1992   11:41AM

Figure 2-3.   DC Meter

- **Function generator,** a graphical window where the user sets up a time varying input to the circuit such as a square wave, pulse train, or one of numerous other standard wave shapes, or else an arbitrary signal drawn with a mouse, or typed in by defining the various points of the waveform.

- **Oscilloscope,** a graphical waveform display tool for transient analysis, which also performs time domain waveform measurements such as rise time, overshoot, settling time, etc. See Figure 2-5.

- **DC voltmeter,** which displays operating point information and specific DC calculations is shown in Figure 2-3.

- **Frequency sweeper,** which defines the frequency range and input for an AC analysis.

- **Network analyzer,** which displays frequency domain information such as gain and phase, formats graphs such as Nyquist and Nichols plots, which is used for calculations such as bandwidth and phase margin. See Figure 2-4.

- **Spectrum analyzer,** which performs Fourier analyses of any waveform from the Oscilloscope tool.

    The Analog Workbench also contains a number of other tools which do not have an exact laboratory counterpart. The most commonly used include:

- **Parametric plotting,** which allows users to run "what-if" analyses on a given design. This is primarily used for analyzing design trade-offs, and 'tuning'. With Parametric Plotting, the user defines a range of

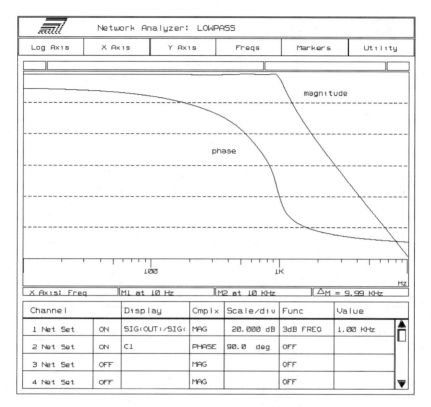

Figure 2-4. Network Analyzer

possible values for a circuit element or parameter. The system automatically sweeps the parameter and records the changes in the simulation results. These are then graphed. For example, you might use the tool to see how performance will change over different values of temperature. Or, you might look at how the start-up time is affected by load. The tool allows users to setup these kinds of analyses quickly and easily. See Figure 2-6.

- **Monte Carlo Analysis,** a statistical analysis which simulates the production environment and predicts the production yield of a new design, based on the cumulative effect of component tolerances. Each component or parameter is assigned a tolerance. The user then instructs the tool to run a large number of repeated simulations with random component values chosen within the tolerance ranges, and records the differences in the simulation results. Yields are calculated automatically based on

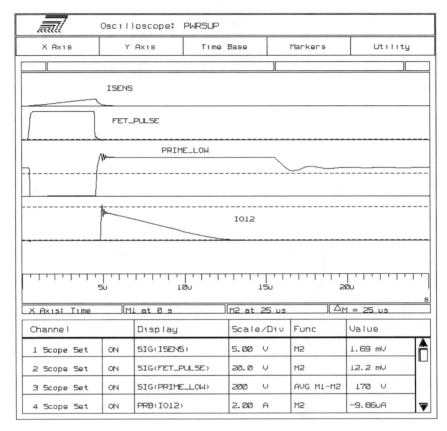

Figure 2-5. Oscilloscope

user specified design limits, and the results are then displayed graphically as shown in Figures 2-7 and 8.

- **Sensitivity analysis**, used both to optimize performance and to test manufacturing of a new design. This tool, as shown is Figure 2-9, tests to see which individual components or parameters impact a design's performance the most. For example, suppose the Monte Carlo tool has identified a yield problem with ripple; the sensitivity tool can now be used to identify which components have the largest impact on ripple, in order to help the engineer fix the problem. In addition, Sensitivity Analysis is often used as a gross manufacturability check. Suppose it is found that a spacecraft supply has a high sensitivity to temperature variation; since a spacecraft can be expected to experience a wide variety of temperatures, this could be a real problem. It is very desirable to

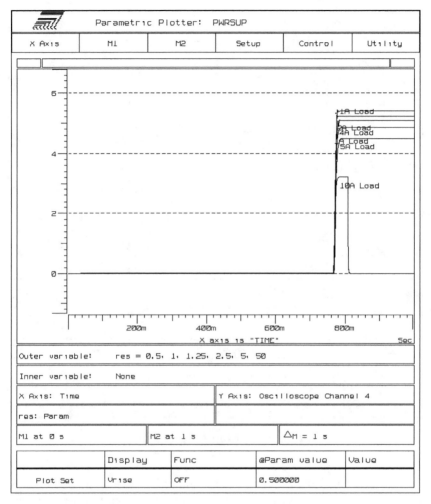

Figure 2-6. Parametric Plotter

discover this early in the development cycle. The Sensitivity tool helps engineers to do this.

Stress analysis or **smoke alarm** is a manufacturability and reliability tool, shown in Figure 2-10, used to make sure that a design will not fail because components are being pushed beyond their safe operating limits. Power dissipation in an analog component is very dependent on the circuit. The Stress Analysis tool first calculates the Safe Operating Area for each compo-

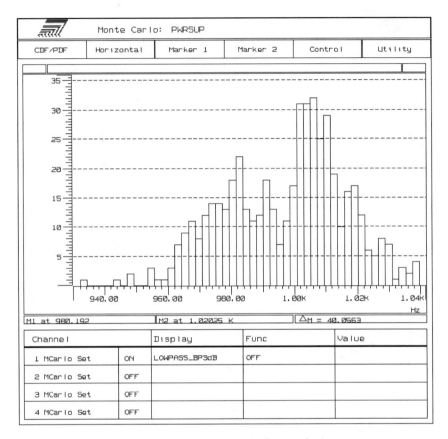

Figure 2-7. Monte Carlo Analysis

nent in the design, then examines the simulation output to determine how close each component came during the simulated operation to exceeding the Safe Operating Area. A variety of reports are produced, and components which actually exceeded their de-rated maximums are highlighted on the schematic.

Testing During Simulation

The goal of simulation is to provide the designer with a circuit that will operate as expected when it is first physically realized. For more complex systems such as surface mount, hybridization, custom semiconductors and integrated circuits, the simulation stage is becoming the breadboard stage with first physical products now being made available at the prototype stage.

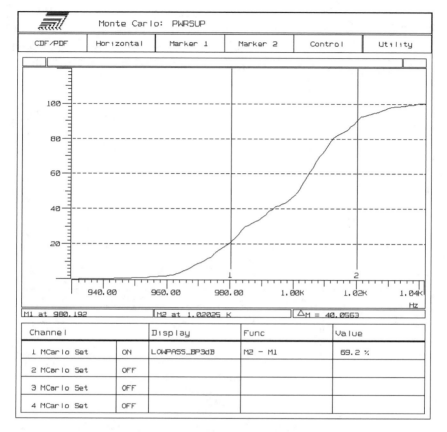

Figure 2-8. Monte Carlo Analysis II

Simulation provides the designer a method of testing out his or her design and circuits before a single component is purchased. Simulation also can predict which part tolerances will affect the production of product, allowing the designer the freedom to design product that will provide high yields on the manufacturing floor, reducing labor and materials costs. Let's take a closer look at testing power supplies during simulation phase of a design and how comparison of test results with the simulation results, from the perspective of the design engineer, can be useful. Testing during simulation is really a design verification process. Additionally, simulation may include the physical test system and interconnect of a power supply under test, to better predict the results of test fixtures and test systems.

Most simulation tools allow the user to determine performance of circuits and components under specified conditions. What is required is an analysis

Sensitivity/Worst Case Analysis: PWRSUP						
Display		Order		Control		Utility

Component	Parameter	Nom	Min	Max	Relative	Sensitivity
Top						
C5		910p	1.001n	819p	-3.361	
R5		20K	22K	18K	-2.853	
R4		20K	22K	18K	-2.831	
C4		75n	82.5n	67.5n	-2.331	
C3		2.4n	2.64n	2.16n	-1.950	
R3		20K	22K	18K	-1.798	
C2		54n	59.4n	48.6n	-1.613	
R2		20K	22K	18K	-1.318	
R1		20K	22K	18K	-1.110	
C1		26n	28.6n	23.4n	-724.731m	

Channel		Display	Min	Nom	Max
1 Sens Set	ON	lp_bw_var	828.992 Hz	1.002 KHz	1.236 KHz
2 Sens Set	OFF				
3 Sens Set	OFF				
4 Sens Set	OFF				

Figure 2-9. Sensitivity Analysis

of all components to determine that, under worst case conditions, components do not exceed their ratings. This is a requirement to provide manufacturing with a design that, when assembled, will function provided that no errors occurred in the manufacturing process. The list of simulated components includes: resistors, capacitors, inductors, magnetic cores, diodes, zener diodes, transistors, analog and digital ICs, power controller ICs (with models of both detailed controllers and state-average controllers, for each type), hybrids, transformers, nonlinear magnetic cores, fluorescent lamps, varistors[8], transzorbs, and glo-bar resistors.

Any simulation is only as good as the models of the components used. The more complex the circuitry, the more important the detail in the models and the level of SPICE used. Some vendors use a partial or reduced set of SPICE designed for PCs, which is fine on small simple circuitry, but on larger and more complex circuits the error buildup can lead the designer to a false conclusion. Designers and their management tend to jump to the

[8] For varistor simulation see application note available from Cadence.

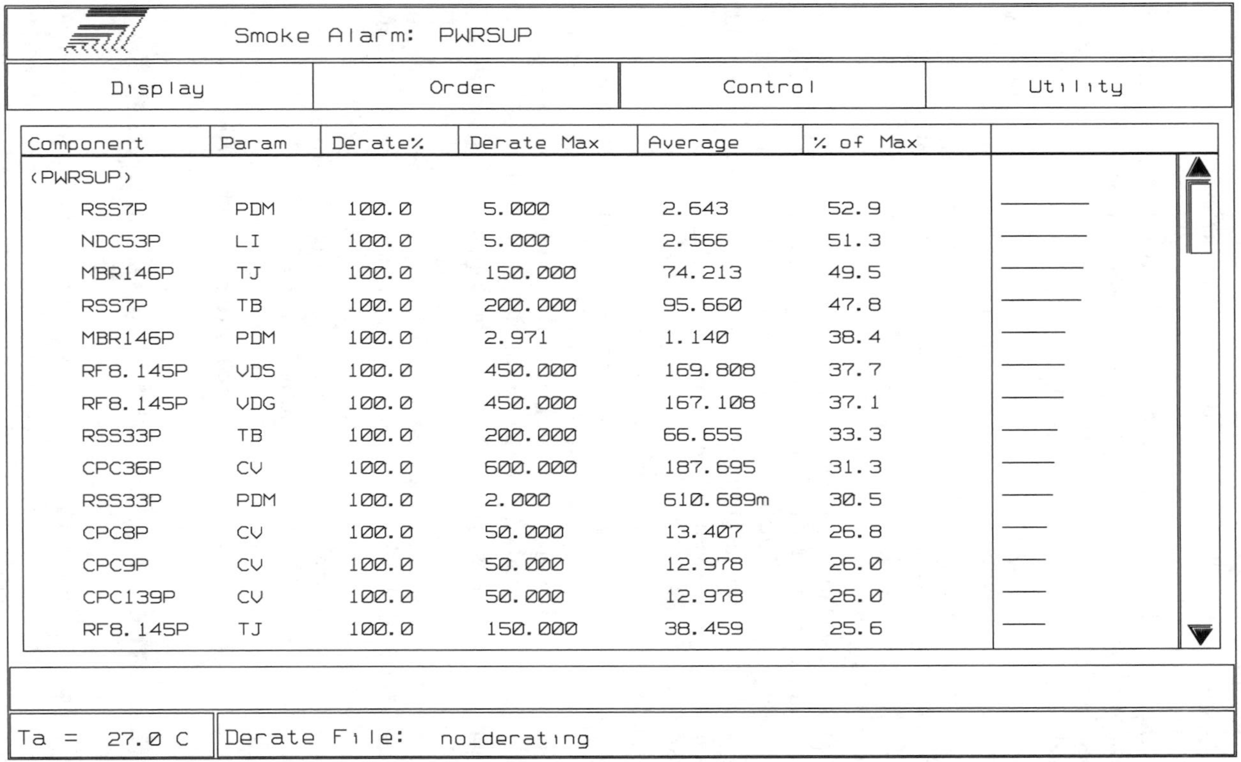

Smoke Alarm: PWRSUP

| Display | Order | Control | Utility |

Component	Param	Derate%	Derate Max	Average	% of Max
<PWRSUP>					
RSS7P	PDM	100.0	5.000	2.643	52.9
NDC53P	LI	100.0	5.000	2.566	51.3
MBR146P	TJ	100.0	150.000	74.213	49.5
RSS7P	TB	100.0	200.000	95.660	47.8
MBR146P	PDM	100.0	2.971	1.140	38.4
RF8.145P	VDS	100.0	450.000	169.808	37.7
RF8.145P	VDG	100.0	450.000	167.108	37.1
RSS33P	TB	100.0	200.000	66.655	33.3
CPC36P	CV	100.0	600.000	187.695	31.3
RSS33P	PDM	100.0	2.000	610.689m	30.5
CPC8P	CV	100.0	50.000	13.407	26.8
CPC9P	CV	100.0	50.000	12.978	26.0
CPC139P	CV	100.0	50.000	12.978	26.0
RF8.145P	TJ	100.0	150.000	38.459	25.6

Ta = 27.0 C | Derate File: no_derating

Figure 2-10. Stress Analysis or Smoke Alarm

conclusion that all versions of SPICE are created equal. Nothing could be farther from the truth. If you are going to use something less than full SPICE PLUS, then make sure that you clearly understand the difference that will result in your product.

The schematic editor used in the Analog Workbench is called GED. It was developed by Valid for some of its earlier products. When the Analog Workbench product was purchased, one of the transitions that was made was to the standard Valid editor which is GED. This version of GED is much more user friendly than previous versions. It allows a hierarchical structure so that a drawing can contain a number of subdrawings for subassemblies, custom hybrid ICs and custom ICs. It has most of the features that you expect in a schematic capture tool. For more info on the Analog Workbench, contact Cadence at: http://www.cadence.com.

The use of worst-case evaluation tools enable the designer to ensure that products designed will be manufacturable with a high degree of success possible with the choice of parts and part tolerances that make up the design. Without worst case design, some of the manufactured product will fail testing, while all components meet specifications. This is an intolerable condition that will lead to poor volume, high costs, poor quality, and eventually poor morale in the production test and assembly workforce.

The Analog Workbench simulation can be check-point and restarted at some point in time after 0.0 as long as the topology of the circuit does not change. Value changes are permitted. This can be a real help when the simulation takes a long time to reach the stable state at first power-on. Once the outputs are stable, fine tuning the design by making changes in component values and using the scheme of check-pointing the simulation, can save a lot of time.

Some simulation tools have the ability to simulate the performance of assemblies of various levels under worst case or other specified conditions, thereby proving that purchased good components, when assembled correctly, will function just as expected. This is the perspective from which we will examine testing with simulation tools. The Analog Workbench allows the designer to recall a design that he or she has been working on, and not only brings up the schematic but also the instruments and their settings.

Simulation of Switching Power Supplies

Simulating switched-mode power converters involves some unique obstacles. One is to acquire models of sufficient accuracy. The Berkeley SPICE program was originally intended for the small transistors of ICs, not for the power MOSFETs and other devices used in power supplies. As a result, these models must be added by a software vendor who specialized in power electronics simulation. The same is true of the transformer. In order to simulate accurately the switching characteristics of the supply, the hysteresis and saturation characteristics of the transformer must be modeled. In addition, the model must not cause "convergence" problems. One characteristic of analog circuit simulators is that sometimes they calculate a point in the simulation. When this happens, the simulation terminates and delivers the user a so-called "convergence error" message. Many convergence difficulties are caused by models, and nonlinear magnetics models are a particularly notorious source.

A second problem with simulating switching power circuits is the "multirate" problem. Analog simulators in general have trouble with circuits in which there are signals with a widely differing time constants. In order to maintain accuracy on a high-frequency signal, SPICE must evaluate very small increments of simulated time, thus taking very large amounts of computation to simulate any low frequency signals that might be present in the circuit

Consider a 250 kHz switcher, for which we would like to see 100 ms of response to a transient. In order to simulate the entire 100 ms, the simulator must analyze 25,000 cycles of the switching cycle. That would take not only a lot of simulation time, but a lot of disk space to store the results on!

A few other common multirate problems are phase locked loops, sigma-delta modulators, and switched capacitor filters, just to name a few. Multirate circuits are generally difficult to simulate unless special behavioral modeling techniques, or even custom simulators can be found. One example is SWTICHCAP, a simulator for switched capacitor filters. For power supplies, a simulator called State Space Averaging modeling was created.

Simulation of Motor Controllers

The Analog Work Bench simulation tool is very flexible and performs well testing systems response tests on power systems such as motor control systems. Factors such as torque, inertia, back-emf, viscosity or, in fact, anything that you can write an equation for, are there for the designer to include in the simulation of the total system design.

This tool can also simulate a system design where function blocks are used in place of schematics that have yet to be designed. This allows a designer to check each subsystem of the design without having to wait for the whole system design to be completed. This can be very useful where custom components with very long lead times cannot be fully characterized. The specification of the part can be used for initial characterization with prototype and production parts characterizations being used as they become available.

An entire system can be simulated with characterization of each functional block, before one part is purchased. With very large power supplies, or motor controllers, this capability can save much time and money, and helps to get your product to market sooner.

Modeling

The first step in simulating supplies is to obtain the necessary models. The easiest way to do this is to buy them from your vendor. It may be expensive, but it will be worth it, especially for parts used in several different projects or designs. Cadence and some other Electronic Design Automation (EDA) software companies have large model libraries with models in use at many, many customer sites. Cadence in particular, supplies models for most of the common Micro Linear and other industry standard controller ICs, so you will not have to spend months of work doing the models yourself. Pretty much the same is true of any power discretes that are used, except that the modeling effort is smaller than that of doing a controller.

If you decide to do your own modeling of semiconductors, Antognetti and Massobrio's book is a good reference for modeling discretes. Access to a curve tracer will be essential, as many manufacturers' data sheets do not have all the information necessary to construct an accurate model.

If you are modeling controllers, an analog behavioral modeling language is a very useful tool. Cadence offers one called Profile, which is both powerful and easy to learn. A model written in Profile simulates in SPICE PLUS along with any other circuitry around it. Unlike discretes, you can usually get all the information you need from the standard IC databooks.

To achieve extremely high levels of accuracy in a switcher simulation, it may be necessary to construct detailed models for the passive components; namely, resistors, capacitors, especially magnetic cores and windings. Michelet and Osman, in an application note, describe in detail how to model all of the passives in a supply using an HP4194 Impedance/Gain-Phase Analyzer.

Once models are available for the individual elements in a switching supply, the next stage is to model the supply itself. This is done by entering the design on an Analog Workbench and simulating several cycles of operation. Once an acceptable design is reached, the real use of simulation begins.

Detailed Simulation

Computer analysis tools have reached a fairly high level of sophistication, and the engineer has a fairly broad array of tools with which to work, including the following.

- **Sensitivity analysis,** which identifies which individual elements within a design impacts its performance most
- **Monte Carlo analysis,** which simulates the cumulative effect of component tolerances on a design
- **Stress analysis,** which was discussed earlier
- **Pole-Zero analysis** for examining loop characteristics
- **Parametric analysis,** which graphs the impact of individual component or value changes on the overall performance
- **Reliability analysis,** which calculates MTBF

The use of these kinds of tools enables a development team to achieve a very high confidence level that a new product will work correctly on release, and that the test methodology developed along with the actual supply will detect any undesirable conditions or problems.

It is a good idea to maintain a current breadboard along with the simulation. There will always be some level of uncertainty associated with the modeling. With care, this can be made to be very small. It cannot yet be eliminated completely. A working breadboard, which produces the same answers as the computer simulation, is the surest way for an engineer and his or her management to remain comfortable that the design is completely realistic. Many people actually employ a dual process, in which the experimentation and what-if analysis is done on the simulation, but periodically the breadboard is brought up-to-date and verified. This kind of careful attention, using both traditional methods and computer analysis together, has reduced the many prototype iterations, design flaws, field recalls, and in-production tweaking to near nonexistence.

Detailed, cycle-by-cycle simulation is very useful for examining characteristics associated with switching, such as supply efficiency, ripple, component stress levels, temperature rise, switching times, power factor, and others. In addition, the impact of layout parasitics can be predicted by adding small passive components to represent them in the schematic, and in the simulation.

A very useful innovation for power supply simulation, one that has made detailed level simulation much more convenient, is the ability to set simulation checkpoints. This allows the user to start a new simulation from any point in a previous simulation, instead of from time zero.

Checkpoint/Restart is commonly used to analyze a supply in steady state. The user runs one simulation from time zero, and the simulated supply starts at zero and turns on over many switching cycles. Once the steady state is reached, the engineer saves a checkpoint. The next time simulation is run, perhaps even with minor circuit changes, the user may start from the checkpoint already in steady sate, and not from time zero. This way the startup period and much of the simulation time can be avoided.

State-Space Average Simulation

The biggest problems with detailed simulation generally relate to the time required to simulate large numbers of cycles. One example is the examination of a power supply's startup time. An inconvenience is the inability to analyze a switcher in the frequency domain, to measure the phase margin for an entire supply, for example. The latter is understandable. Switching power supplies are extremely nonlinear, and aside from a few exotic tools, all analog simulation in the frequency domain is linearized.

There is a technique called state-space average modeling which solves both of these problems. This concept was originally described mathematically by Dr. R. D. Middlebrook of Caltech. State-space averaging was first applied to SPICE by Dr. Vince Bello of Norden Systems. Since that time, it has been considerably refined by Bello and others, including Jeff Berwick and Dr. Mehrdad Tabrize of Cadence.

The state-space model solves the multirate problem by eliminating the switching waveform completely. Instead, the supply is considered always to be a linear combination of its two states, with the weightings of the two states determined by duty cycle, a new variable. By modulating the duty cycle variable, a model of the entire closed-loop supply can be produced.

To build a complete state-space average model for the power supply, the controller and transformer sections are replaced with new state-space average versions. The rest of the circuitry stays the same.

The simulator no longer need calculate all the short-term cycles in order to derive the long-term behavior. A state-space average simulation runs about one hundred times faster than an equivalent detailed simulation, allowing not only simulation of long-term effects such as start-up and response to a transient, but also even longer analyses like Monte Carlo and parametric analyses, in which many simulations of the same circuit must be sequenced.

In addition, state-space average simulation has the characteristic that it is a linear analysis, and can therefore be used in the frequency domain. It is therefore possible to make direct measurements on frequency domain parameters such as phase margin and gain margin.

Two drawbacks of state-space average simulations are that no real information can be obtained about switching related characteristics such as efficiency and that a different physical schematic is usually required for the state-space average model.

Modeling the Effects of Layout Parasitics

Layout parasitics can torpedo a design. They are usually most significant in high-performance, high-precision, and power supply designs. They can be particularly troublesome because they can be very hard to estimate, and they usually appear in the development cycle after most of the electrical design choices have already been made.

Layout parasitics of many kinds can affect power supplies, but inductance is typically one of the most troublesome, due to the fast-switching currents which abound in switching power supplies. Other parasitics can include unexpected capacitance's at high frequencies, undesirable thermal effects, and resistances.

Most printed circuit board layout is done on computers today, so it is curious that so few layout software packages calculate these kinds of parasitics, at least as of this writing in 1997. One of the few which does this, is Cadence's Allegro software, which has the added benefit of interfacing seamlessly with the Analog Workbench. In addition to a variety of digital parasitics, Allegro calculates capacitances, inductances, resistances, transmission line parameters, and thermal effects.

Thermal analysis of an analog board is one of the most interesting applications of computer simulation using a variety of techniques and tools together. In order to develop a thermal profile of a printed circuit board, two crucial pieces of information are needed: the actual placement of components with respect to each other, and the average power dissipation of those components, along with certain package information. Most conventional thermal profile tools were designed for digital circuits, and taken into account only the placement, with the power dissipation being obtained from a table or library. This is acceptable for most digital circuits. For example, the power dissipation of a 74LS00 TTL gate is about 4 mW, irrespective of the rest of the circuit it happens to be in (although it does vary slightly with clock frequency).

Analog printed circuit boards, such as power supplies, are much different. Although the physical placement information is still critical, the power information for the components can no longer be assumed. Power information is dependent on the actual circuit. The average power dissipation can only be obtained from simulation (or a breadboard). Furthermore, SPICE does not calculate the information directly, it must be obtained from a post-processing step.

Fortunately, the Analog Workbench stress analysis tool (smoke alarm) performs all the postprocessing to calculate this information, and then passes it to Allegro. Allegro then calculates the junction temperature for each element, based on the power information and the component placement. Once the junction temperatures are available, they can be back annotated directly to the Analog Workbench schematic for resimulation. In order to complete the loop, Cadence modified SPICE PLUS to allow different temperatures on different elements within the same design. A few iterations may be necessary to achieve full accuracy (the temperature changes may cause the power dissipation's to change), however, the result is the ability to predict operating temperatures across a power supply board as well as their impact on overall electrical performance and reliability.

Simulating Specific Measurements

Once the design is essentially complete, the state-space average and detailed simulations calibrated to each other and to the breadboard, and the most important layout parasitics incorporated, the process of characterizing the design for testing can be completed.

Test Simulation

Power supply testing is performed to measure the performance characteristics of a unit under test. To simulate this testing we are actually simulating the operation of the product. The only part that needs to be added is the affect that the tester has on the unit under test. Added capacitance and inductance in cables are a major consideration. In this case it is not uncommon to discover a design fault that may not be readily visible in benchtop testing or even in a systems test, due to masking by either the bench setup or even the application itself.

Advanced Test Development

Simulation of new test scenarios, ideas, and concepts without the expense of implementation in hardware, dollars, and time can often be accomplished with good results. For example, simulation of a printed circuit card in a test set designed for it can provide the data necessary for designing the card test system interface. This can be quite important in systems where probes can load down some sensitive points of measurement in a circuit. Built in self-test circuits can be simulated along with the system design to ensure that no unwanted interactions will result.

Final Simulation

Final simulation can be described as the addition of circuit board parasitics to the simulation, to ensure that the design is still manufacturable in the real world. This process takes place as the Printed Circuit Board design is completed, but before it is finalized. This is especially important in designs approaching or above 100 KHz, and is critical in designs of 1 MHz or more. While this process has yet to be automated in a commercial tool that we are aware of, the process can still be effective by selecting those circuit elements where parasitics can have either a direct effect or a causal effect, or both. These parasitics are then added to the schematic, and the simulation rerun to determine the exact effect that they will produce on the design.

Military Use of Simulation for Test

One military depot reportedly uses the Analog Workbench to simulate power supplies as a means of obtaining waveforms for test points in the unit at various operating points. These waveforms are used for troubleshooting failed units which are repaired and returned to service. Because the large number of power supplies that they service are made up of many, many different part numbers, are of many different types of power supplies, training for personnel is difficult. Qualified technicians can troubleshoot a power supply from simulated waveforms with excellent results. The savings in training and the cost of otherwise obtaining good units and making measurements is prohibitive.

The Cost of Simulation

The cost of simulation is a complex issue. Cost considerations must include:

- Cost of software
- Cost of hardware
- Cost of added training
- Added income due to a shorter time to market
- Reduced costs of servicing a better designed product, and
- Cost of increased overhead

The biggest item is the additional income owing to time to market changes for designing reliable products. Once you understand the time impact to the project, marketing and finance personnel can help you determine the worth to the company to get the product to the market in the shorter time. Shortening the development time for products can allow your marketing folks to seek business that they would have otherwise ignored. Time to market costs can easily run into millions of dollars on some projects.

Table 2-1. SPICE Performance Tests

Test	Initial Operating Point	Tool and Model	Approach
Stress/Temp	line/load cond.	Smoke Alarm, Detailed model	Starting at a Steady State for a given line & load condition, run the Smoke Alarm tool over a couple of cycles and record the RMS, Average or peak stress on any component. You can select a derating criteria for all of your components and determine the components which fail. Select a new steady state point and repeat the process.
Output Voltage Distribution	line/load cond.	Sensitivity/Worst case analysis, Parametric Plotter State Average	Define tolerances on all of the components in your circuit. Perform the sensitivity analysis to identify the critical component(s) that affects the Output voltage. Then using the Parametric Plotter, you can see the family of curves for the output voltage variations
P-P ripple dist.	line/load cond.	Parametric Plotter, State Average	Identify the parameters to vary. Then vary the parameters within the spec limit by inputing the start and end times and the number of steps you wish to take. This will provide you a family of curves for the ripple voltage for the different line/load combinations.
Phase/Gain dist.	line/load cond.	Parametric Plotter, Network Analyzer State Average	Identify the parameters to vary. Then vary the parameters within the spec limit by inputing the start and end times and the numberof steps you wish to take. This will provide you a family of curves for the phase/gain for the different line/load combinations.

continued

Table 2-1. *Continued*

Test	Initial Operating Point	Tool and Model	Approach
Turn-on charac.	line/load cond.	Parametric Plotter, Oscilloscope State Average	Identify the parameters to vary. Then vary the parameters within the spec limit by inputting the start and end times and the number of steps you wish to take. This will provide you a family of curves for the turn-on charac. for the different line/load combinations.
Turn-off charac.	line/load cond.	Parametric Plotter, Oscilloscope State Average	Identify the parameters to vary. Then vary the parameters within the spec limit by inputting the start and end times and the number of steps you wish to take. This will provide you a family of curves for the turn-off charac. for the different line/load combinations.
Transient Response	line/load cond.	Oscilloscope, Function Generator, State Average	Input an input stimuli in the Function Generator and monitor the response in the Oscilloscope
OverVoltage Prot.	line/load cond.	Oscilloscope, Function Generator, State Average	Input an input stimuli in the Function Generator and monitor the response in the Oscilloscope
Over Current Prot	line/load cond.	Oscilloscope, Function Generator, State Average	Input an input stimuli in the Function Generator and monitor the response in the Oscilloscope

Other Tools

One of the most popular software tools is the program PSPICE. It is very popular due to the number of platforms that it will run on including IBM-PC and clones. PSPICE has been evolving and is starting to catch up to the really sophisticated tools such as the Analog Workbench, by adding more capabilities and increasing the number of models in its libraries. Even so, some believe that the true value of tools running full SPICE or SPICE PLUS is in the accuracy of the results. This accuracy becomes more obvious as the size and complexity of circuit simulations increases. Model accuracy is also an issue that is bandied about by those playing in this arena. You must be the one to make the best decision based on your design requirements and funds available. Some would have us believe that it is purely a cost justification issue, once you remove the emotions of the engineers involved. In all actuality, it really is some of each.

Bibliography

1. Antognetti, Paolo, Guiseppe, Massobrio. *Semiconductor Device Modeling with SPICE,* McGraw Hill, New York. 1988.
2. Bray, Derek. "Improve Differential Amplifier's Linearity and Dynamic Range," *Electronic Design,* March 3, 1987.
3. Carter, Don. "Better Software for Analog Simulation," *Marketing Manager,* January 1987.
4. Comeford, Richard. "Circuit Simulation Opens a New Frontier Thermal Effects," Contributing Technical Editor, *Test and Measurement World,* January 1989.
5. Filseth, Eric. "IC Test Applications of Analog CAE," presented at the ATE (Automated Test Equipment)/West Conference, January 1989.
6. Helliwell, Kim. "An Improved Voltage/Current Controlled Switch Model," presented at the 23rd Inter-Society Energy Conversion Engineering Conference, July 1988.
7. Jaycox, Jeffrey. "CAE tools break barriers of feedback measurement," *Electronic Design,* May 28, 1987.
8. McGregor, Jim; Hal Conklin. "Analyzing Manufacturability and the Effects of Design Changes," *Printed Circuit Design,* Vol. 3, No. 5, May 1986.
9. Michelet, Robert. "Advantages of Using Real B-H Loops in Power Supply SPICE Simulations," *Power Conversion and Intelligent Motion,* August 1987.
10. Michelet, Robert. "Evaluating the Half Bridge Inverter Circuit at High Frequencies with CAE Tools," presented at the High Frequency Power Conversion Conference, April 1987.

11. Michelet, Robert. "Using CAE Tools to Evaluate the Inverter Section of the Half Bridge Switcher," presented at the Power Electronics Conference, April 1987.

12. Orgain, Cheryl. "Analog Design Tools Trip ASIC Trouble," *Electronic Engineering Times,* November 23, 1987.

13. Orgain, Cheryl. "Create Linear ASIC Macros Without Spice Nightmares," *Electronic Design,* December 10, 1987.

14. Tabrize, Mehrdad; Cheryl Orgain. "Analog CAE Tools Go Beyond Breadboarding," *Power Conversion and Intelligent Motion,* January 1988.

15. Tabrizi, Hehrdad; Martin Walker. "Computer Aided Engineering for Power Electronics using Spice Plus," presented at the Automated Design and Engineering for Electronics Conference, April 1987.

16. Tabrizi, Mehrdad. "Nonlinear Magnetic Model Realistically Simulates Core Behavior," *Power Techniques,* March 1988.

17. Walker, Martin. "Analog CAE Workstations," presented February, 1987 at the Systems Design and Integration Conference.

18. Walker, Martin. "Analog IC Design-to-Test Integration," presented at the ATE West Conference, January 1989.

19. Walker, Martin; Eric Filseth. "Analog IC Simulation for Manufacturability," *VLSI Systems Design,* January 1988.

20. Walker, Norman; Martin Walker. "Design Current-Mode Switching Supply on Analog Workstation," *EDN,* June 1986.

3

Functional Testing

AC Input Tests

Power supplies are generally powered by either AC or DC power, and most are powered by the AC line at a frequency of 50, 60, or 400 Hz. The first tests are performed to check that the input is not shorted and will not draw excessive current when testing begins.

NOTE
These tests are not presented in any particular order. Do not impart any importance to the order in which they are discussed.

First Power-Up

The first time that AC power is applied to a particular unit under test, it must be applied in small steps while monitoring the AC line current to detect shorts and other assembly faults. Typically 1.0 V is applied to look for hard shorts, then 10.0 V is applied to look for shorts or excessive current conditions. The voltage is then stepped in larger increments to full power-up, at nominal line conditions. If an excessive current condition is detected during this test, power is removed and further testing is then performed to isolate the problem.

Inrush Current

Accurately testing inrush current requires that the impedance of the AC source providing power to the unit under test be as low as any source that will ever power the unit. If the source impedance is higher during testing, then a lower peak inrush current reading will result. Typically inrush current is measured at either 0° or 90°, as determined by the applicable specifications. When the phase angle is not specified, the 90° worst case scenario is usually assumed. This test is performed by first discharging all capacitors in the unit to be tested. Then the specified maximum-load specified is applied to all outputs per the specifications. The AC power is then switched on at the specified voltage, frequency, and phase angle, while the inrush current draw of the unit under test is recorded. Particular care must be taken in either specifying the impedance of the AC line, the source of the AC power. Power lines in residential neighborhoods tend to have higher line impedances, generally falling between 0.03 and 0.10 Ω with fault currents in the region of 1,000–5,000 A. In industrial areas and office complexes, fault currents as high as 200,000 A may be seen.

Softstart

Softstart circuitry is used to reduce inrush current ratings for power supplies. This is especially important in power supplies where the total power drawn from the AC line approaches the maximum rating of either the power cord or the branch circuit into which it is plugged. The Softstart test is used to provide that any softstart circuitry designed into the power supply is operational. Several techniques may be used for softstarting circuitry, including the following:

Thermistors (Glo-Bars)

Thermistors are slow devices, and very fast repetitive cycling of the power can cause surges that may be harmful to the unit under test. Plan the testing of thermistor protected power supplies carefully.

Resistors and Relays

After some time delay, the relay closes, allowing full current to the input section of the power supply. Before the delay passes, current is limited by a power resistor in series with the AC line. Problems can arise in this circuit due to relay contact rating problems, as well as possible dust and dirt contamination of the relay contacts.

Resistors with Either an SCR or a Triac

In this scheme, a solid state device replaces the relay. The solid state device, either an Silicon Controlled Rectifier (SCR) or a Triac, replaces the relay in the circuit described above. One obvious difference between this type of current limiter and the relay is that the forward voltage drop, of the semiconductor, is added in series with the line. In testing all these devices, the concern is to measure the indication that the devices were first in the open condition, and then closed to allow operation at full loads.

High Line

High-line testing is worst at the lightest load condition specified. The purpose is to test the power supply while the input section is under the greatest voltage stress. All specified outputs are tested at this input condition. Alternating line current needs to be monitored for faults.

Low Line

Low-line testing is the worst case test made at full load. The input rectifiers, diodes, and capacitors feel maximum continuous stress under these conditions, as peak current reaches its maximum value. Specifications may sometimes specify the voltage for this test to include a lower value for brownout protection.

Static Line Regulation

The output of the power supply under test is measured at the various in voltages that cover the range of input power for the unit. Because this is a static test, the output is allowed to stabilize after each input adjustment.

Dynamic Line Regulation

Dynamic line regulation is a measure of the power supply to maintain all outputs within specification during changes of the input power. The specifications must define the changes to the input power and the rate at which those changes occur. In lieu of any specific requirements for dynamic line regulation, the input will slew (step) from nominal line voltage to high-line to low-line to nominal, and again from nominal line voltage to low-line to high-line to nominal line voltage. In today's dirty electrical environment this can be an important specification.

Power Factor

Measurement of the power factor for a power supply is becoming an increasingly important task. This is especially true where power factor correction is designed into the power supply. The power factor is the ratio of the true power to apparent power consumed by the power supply. This is expressed as a decimal, and is identified as leading or lagging the voltage. The symbol θ is used to represent the phase angel between the voltage and current, assuming that both waveforms are identical in shape. Several methods may be used to measure the power factor. Occasionally, a specification will indicate which method is to be used.

Start-Up Time

The start-up time or start-up delay is the time delay between the application of input power and the time at which all outputs are within specification and stable. Where applicable, this can also apply to remote control inputs of a power supply as well.

Hold-Up Time

Hold-up time is an important power supply specification. Hold-up time is the time during which a power supply's output(s) remain within the specified limits after the loss of input power. It is sometimes defined as the maximum length of time that the input power can be removed for without losing regulation at the outputs. In both cases, the test is made under worst-case conditions of low line and maximum load on the output(s). It is also called the ride-through test.

Leakage Current

Leakage current is one measure of safety considerations in power supply design. The leakage current is the measured AC or DC current flowing between input and output of an isolated power supply. In some cases the chassis or ground leads are connected to the output. This provides a measure of the current available should a ground connection be broken and a person or equipment be exposed between ground and the power supply outputs and/or chassis. This test is usually performed at some specified voltage. In medical equipment, especially patient connected equipment, Underwriters' Laboratories (UL) specifies the current limits and the testing procedures or methods in UL-544. Equipment for other markets may require testing in

UL or other regulatories publications, regulations, or standards. Two areas of concern here are usually the AC line filter current, and the leakage current in any transformers. Transformers, even very large ones, can be constructed to exhibit very low leakage currents.

Grounding Tests

Grounding tests determine that all of the grounding connections and chassis connections are properly connected and exhibit a resistance below some specified limit.

Ground Current

Ground current is a measure of the current flowing in the ground pin of the power cord, or other appropriate connection point to ground.

Common Grounds

On some power supplies there are more than one ground. Depending upon the application, these grounds may be connected by a specified resistance, and local or remote sense connections made to each of them. The test generally performed is a resistance check between the two grounds as a short could cause excessive or unwanted currents to flow in these circuits.

Isolated Grounds

Some outputs of some multiple output power supplies are isolated from the other outputs. Testing is performed to check the isolation. This testing may be performed during HIPOT testing.

Other Tests (Per Specification)

As almost always, other specifications may need to be tested. Use care in selecting the test that actually checks the specification you intended.

DC Inputs

Direct Current input tests, where they are substantially different from the AC input test, are examined here.

Inrush Current

Inrush current measurement is different in that the only limit to an instantaneous infinite current draw is the impedance of the source, and the impedance of the interconnecting cables and wiring.

Efficiency

Efficiency is a measure of the power consumed in the process of power conversion. For DC-DC converters, it is simply the Watts Out divided by the Watts In, multiplied by 100%. For AC-DC and AC-AC converters, the measurements process requires that careful attention be paid to the RMS measurement of both voltage and current, as well as the phase relationship between them. For some switching power supplies, a good indication of losses can be seen by shorting the output(s) and measuring the input power.

Control Inputs

Many power supplies have inputs for controlling various functions. These inputs must be tested. The level of test will be dependent upon the importance of the control function and specification. Some remote control inputs may seem simple to test at first glance, but careful examination may indicate that more extensive testing is required, depending upon where the test is performed and reasons it is performed.

Power Enable

This input enables or turns on the power supply outputs. It may be specified for either CMOS or TTL levels, or something else, but what are the limits of those levels and how should they be tested? A case where a power supply was designed for TTL control of a remote enable signal, failed in the customer's hands because his test set, built in-house, used a 10 kΩ resistor as a pull up. Once the culprit was identified, everyone was happy. When homemade testers are to be used on a new or custom product, it will be to your advantage to borrow one, to check out the products operation on that tester. If it is too large to ship, take a prototype of the power supply to the tester to look for compatibility problems.

Power Disable

Power disable is just the opposite of power enable.

Master Clock Input

On some modular switching power supplies, the master clock is fed to each module to control EMI and RFI. This input, and its ability to control the switching frequency must be tested. Additionally, when this signal is lost, the default condition defined in the specification must also be tested.

DC Output Functional Tests

Static Load Regulation

The output of the power supply under test is measured at the nominal input voltage and various output currents that cover the range for that unit. Because this is a static test, the output is allowed to stabilize after each load adjustment.

Dynamic Load Regulation

Dynamic load regulation is a measure of the power supply's ability to maintain all outputs within regulation to specifications during changes of load current. The specifications must define the changes to the load current and the rate at which those changes occur. In lieu of any specific requirements for dynamic line regulation, the input will slew from 20 to 80% of load at 5 A/μs.

Short-Term Drift (Voltage)

This specification is typically only checked to verify a design, and not performed in volume production, unless as part of an ongoing reliability test program.

Long-Term Drift (Voltage)

Long-term drift is typically only checked to verify a design, and not performed in volume production, unless as part of an ongoing reliability test program.

Thermal Drift

The thermal drift test is usually specified for voltage, but actually may be specified for any parameter. This specification is typically only checked to verify a design, and not performed in volume production, unless this is done as part of an on-going reliability test program.

Ripple and Noise

There are several ways to measure ripple and noise, but the most difficult part of the task is the noise measurement. One important task requiring care and precision is making the connection between the unit under test and the measuring equipment. Leads must be kept short, to prevent masking noise in the higher-frequency bands. Ideally, the measurement should be a differential one. Use the specification as a guideline, but clearly document your measurements and setup very carefully. Some lump ripple and noise into a single measurement and call it periodic and random deviation (PARD). PARD is defined[9] as the sum of all ripple and noise components measured over a specified bandwidth, and stated in peak to peak values. Ripple is harmonically related to the input power frequency, or the switching frequency, or both, while noise is purely random in nature. A technique was used by Interpro in three modes to provide a peak to peak noise measurement to measure spikes down to 10 ns in width, a peak-to-peak ripple measurement, and an RMS measurement of the ripple and noise. This capability is provided in the form of a board and some software in their power supply test systems. Another way of measuring the ripple and noise is by placing the power supply into the actual application and then measuring ripple and noise in place. This practice can only be used when a power supply has a single application to consider, and is not considered by most engineers to be a real test of the power supply. Interpro is now owned by Shaffner and may be contacted at http://www.shaffner.com.

Overvoltage

There are basically two methods used for overvoltage protection. One provides a crowbar to short the output when a certain voltage is reached, while in some newer switching power supplies the control circuitry shuts down the power supply when sensing an overvoltage condition. Some power supplies actually contain both types of protection. Each must be tested independently. The typical backpower test for "crowbar" circuitry using a current limited external source is the widest used method today.

[9] See "Handbook of Standardized Terminology for the Power Sources Industry, 2nd Edition" published by the Power Sources Manufacturers Association, 14 Ridgedale Ave. Suite 125, Cedar Knolls, NJ 07927, phone (201) 538-9170.

Isolation

Isolation of DC, AC, noise currents, and EMI/RFI can each be specified and measured separately. Isolation requirements are highly application dependent. Refer to applicable standards and agency requirements identified in the power supply specification.

DC and AC

These measurements are usually referred to as HIPOT tests or as leakage tests. The product specifications will identify which is to be used.

Noise and EMI/RFI

This specification is sometimes used where the power supply is connected to a known dirty AC or DC source. Most often the power supply may contain a ferroresonant transformer to provide high isolation. Ratings to 80 dB are not uncommon. Shielding schemes are also used in less demanding situations, with correspondingly lower ratings. This test is not normally performed in a manufacturing environment. EMI/RFI testing is often farmed out to a testing laboratory where results can be certified.

Sequencing

It is not unusual for a power supply specification to identify an order for voltages to come up in or to go down in. Sequencing was very common at one time due to memory chip technology, but now only seems to be found in special applications. Testing is accomplished via either a multichannel oscilloscope with a delayed sweep, or more frequently today by an automated test system. Programming is usually a very simple matter once the sequential order is identified.

Cross-Regulation

Cross-regulation is the percentage of voltage change in an output caused by a change in load of a different output. Cross-regulation can only be measured in multiple output power supplies. The power supply specifications will dictate what the limits for cross regulation are.

Current Limit and Short Circuit Protection

Many power supplies have outputs that provide some form of current limit or short circuit protection. The most common form is current foldback where after the maximum current is reached further reduction of load impedance causes the current to fold back to a lower value. These values are specified and measurable. Some designers still use adjustable resistors or factory-selected values for setting these parameters in the manufacturing environment. The trend is toward using fixed components and then testing the results.

AC Output Tests

Frequency Stability

Power sources providing one or more AC outputs will specify frequency stability of those outputs. Depending upon the topology used in deriving the AC outputs, you may want to test for frequency stability against a number of variations, including:

- Time
- Temperature changes
- Input voltage changes
- Load current changes
- Changes in other loads
- Line noise or EMI/RFI

Waveform Distortion

Waveform distortion due to changes in load current, reactive loads, or loads with a nonsinusoidal waveform must be measured. Distortion analyzers are available off the shelf for the task. For infrequent use, consider renting an instrument for a short period.

Spectral Purity

Spectral purity is the measurement of the harmonic content of the output waveform. Since harmonics distort a waveform, this is a measurement of waveform distortion.

Current Limit

It is common as some AC sources to have two current limit mechanisms in use at the same time. One is a slow responding limit of the RMS current value. The other is a faster acting but higher limit providing the AC source with loads containing transformers and motors. The specification will identify the timing relationship and current limits that are to be tested.

Ancillary Outputs

Many power sources have ancillary outputs that may require some form of testing. Except for general-purpose power supplies, each of these ancillary outputs is usually highly application-dependent. Carefully examine the specifications.

AC OK

AC OK circuitry must be tested to specification and usually includes a low-line AC input voltage and a response time for the circuit to reaching that point before changing state.

DC OK

While DC OK circuitry must also be tested to specification, it usually includes a test of specified output voltage values with a very fast response time upon sensing a change in conditions and changing the DC OK output state.

Real-Time Line Clock

Real-time clock signals are not uncommon, and may be derived from the AC line very simply, or with very complex circuitry. Some will require additional testing to prove that they are transient insensitive. The power supply specification is your guide for parameter values to be tested, and perhaps even dictate which tests are to be performed.

Remote Control/IEEE Bus Interface

Some power supplies have provisions for remote control, either via an IEEE bus interface, and an Electronics Industry Association (EIA) interface or some other customized interface. Testing of these input/output (I/O) ports to both the specification of the interface, and the ability to control the power supply and provide information to the power supply specification may be required.

Master/Slave Clock Sync Circuits

Testing master/slave clock sync I/O ports where they exist can be accomplished by providing the specified signals and measuring the result, or by measuring the specified output signals.

Regulatory Agencies and EMI/RFI Testing

The major regulatory agencies that will be discussed here include:

- Canadian Standards Association (CSA)
- Federal Communications Commission (FCC)
- Underwriters' Laboratories (UL)
- International Electrotechnical Commission (IEC)
- Verband Deutscher Elektronotechniker (VDE)
- Technische Uberwachungs-Verein Rheinland (TUV)

Many people are confused as to what role each of these regulatory agencies play. Bureaucratic red tape is minimized, for the most part, and each agency has its own procedures to follow with its ever present paperwork. Because time is recognized as a primary consideration in regulatory testing, start as early as possible, and avoid delays. Standards set up by regulatory agencies are for the purpose of both safety and controlling radio frequency (RF) emissions. Safety standards that help ensure safe operation of equipment are concerned with fire hazards, dielectric withstand voltages and conductor spacing to meet those voltages, leakage currents that may be a hazard to people or other equipment, and proper receptacle compatibility for power input and outputs. Additionally, proper product labeling, temperature limits, and transformer isolation are part of the safety considerations. Let's examine each of these agencies in more detail.

UL (or Underwriters Laboratories) is both a standards organization and a testing agency. Additionally, UL's Follow-Up Services Division conducts unannounced factory visits to monitor product conformity to standards. This is done by inspecting products on the production line for compliance with the specifications to which that product was originally tested.

To submit a product to UL for testing, you first send a letter to the nearest

UL regional office requesting an application for testing and providing them with enough information on the product so that they can send you a cost estimate and request for deposit along with the application forms. The cost estimate is determined by the amount of work that they perceive will be required.

CSA (or the Canadian Standards Association) is somewhat similar to the UL in the United States. CSA operates in Canada in a manner similar to UL in the United States. If you are planning to sell products that may eventually end up being sold in Canada either directly or indirectly as part of a larger system, then you must make your design team aware of any applicable CSA standards. The CSA Field Service Group provides a service similar to UL's Follow-Up Service Division in monitoring product conformity to standards.

FCC (the Federal Communications Commission) has chosen to regulate and govern RF radiation limits, conducted RF emission limits, labeling requirements, and instruction manual requirements for electronic products falling into two classes:

- Class A: A computing device that is marketed for use in a commercial, industrial, or business environment.
- Class B: A computing device that is marketed for use in a residential or home environment by the general public, including: calculators, computers, personal computers, electronic games, and digital watches.

Before a Class A or Class B product can be legally marketed in the United States, the FCC requires compliance with the appropriate standard. FCC Docket 20780, Parts 15 and 18 of the Federal Communications Act specifies and governs the limits. While the test results of Class B products that are connected to a television receiver must be forwarded to the FCC for verification at this writing. All other products must be tested but need not be filed.

IEC (or the International Electrotechnical Commission) is a standards setting organization whose standards are incorporated into government standards. One example of this would be IEC 380, which becomes VDE 0806 with an extensive forward added.

VDE which stands for Verband Deutscher Elektronotechniker, is the standard setting agency and a testing agency in Germany. To meet VDE standards such as VDE 0806, certain VDE-approved components may be required. Care must be exercised in selecting those critical components that meet the appropriate standards.

TUV which stands for Technische Uberwachungs-Verin Rheinland, is a testing agency found operating only in Germany, which tests for compliance to VDE and GS standards. Geprüfte Sicherheit (GS) is the mark issued by the German government certifying VDE compliance. TUV has an office in

Mount Kisco, NY in the United States where employees speak the English language.

Testing Services and Consultants

Consulting and testing services, such as those offered by ITS,[10] are available when assistance is required for a regulatory problem or for testing. This firm is one of the better known firms in the business, providing both technical and regulatory assistance as required. They publish a handbook entitled *Handbook of EU/EMC Compliance* for engineers concerned with U.S. and foreign emissions, telecommunication, ESD, and line transient standards. When dealing with more than one regulatory agency, consultants can be especially helpful.

Annotated Bibliography

1. Sugrue, Denis, "Power Supply Testing. An Overview," An application note, Intepro Systems, March 28, 1988. This application note contains an interesting fixture for PARD measurements, including construction details.

2. ————, "FCC Docket 20780, Parts 15 and 18," Federal Communications Act, Washington DC.

3. ————, *Handbook of EU/EMC Compliance,* ITS, Boxborough, MA, Yearly.

4. ————, *Handbook of Standardized Terminology for the Power Sources Industry,* *2nd Edition,* Power Sources Manufacturers Association, 14 Ridgedale Ave. Suite 125, Cedar Knolls, NJ 07927, phone (201) 538-9170, 1995.

5. ————, "Measuring Ripple, Noise During Power-Supply Test," *Electronics Test,* October 1989, p. 74–75.

[10] ITS, 593 Massachusetts Avenue, Boxborough, MA 01719, phone: (508) 263-2662.

4

Qualification

Qualification of components and/or subsystems is a process that many over-
look, or just don't put enough thought into. Component qualification, vendor
selection, is just the tip of the problem, where like the iceberg, 90% of the
problems lie out of sight beneath the surface of the water. Building quality
products means qualification of vendors who will deliver products with consistent quality, and who, if a problem should occur, will put resources in place
in a timely fashion to resolve the problem to everyone's satisfaction.

Quality

Defining Quality

Quality is defined in terms of an applicable failure rate, today expressed in
terms of parts per million (ppm). Terms such as Six Sigma, Zero Defects,
and other, coined by Motorola and other manufacturers attempt to convey
the idea that their products are manufactured or built to certain measurable
quality levels. While many manufacturers claimed quality levels in percentages in the past, today the Japanese have driven those levels to single-digit
numbers of parts per million. The real goal is ZERO DEFECTS. It is less
expensive to build good product and ship it than it is to build less than
perfect product and then be forced to repair, or even scrap some product.
The simple goal of do it right the first time, can be applied in every process
step or substep.

Designing Quality In

Quality can be designed in by creating products that are designed so that every item produced will operate to specifications the first time. This means that worst case design methods must be employed. International Standards Organization (ISO) Standard 9002 is one attempt at helping manufacturers design systems that produce good product all the time. This standard requires a level of documentation where every process step in a factory is fully documented and followed. Any company wishing to do business in Europe in the near future will be required to comply to some level of this standard.

Assuring Quality

Quality is assured by people. People make quality products, when properly motivated. There are many styles of management, and you must find the most appropriate one for your situation. The most important thing to remember about quality is that everyone in the plant must be involved. Everyone from the Chairperson of the Board down to the Assistant Janitor must understand the quality goals and be actively seeking ways to meet and exceed those goals. The goal of constant and neverending improvement is the only way to really achieve quality processes and products.

Measuring Quality

Quality can be measured over time by many different methods, but measuring quality does not in any way assure quality. Quality products are designed as quality products. Better products start on the drawing board at the conception stages and have quality designed into every phase of the product and process design. Designing quality products must be a team effort.

Feedback on Quality

In today's manufacturing processes a process error can cause considerable rework and scrap. The only secret to building quality products is to ensure that feedback on any problem in production is immediately feedback to the point of origin, and steps taken at once to correct the problem. If this process is followed, costs will be reduced and quality will steadily increase. This is the one area where American manufacturing companies continually fail to put enough emphasis, and miss the boat by trying to inspect quality into products.

Burn-In/Environmentally Controlled Stress Testing

Creating a universally acceptable concept of the burn-in process and for the terms used in describing the burn-in process has always been an almost impossible task. Nearly everyone has a different concept of the solution to the problem, and even of the problem of power supply quality itself.

The first step will be in defining the terms used in discussing the process, for common terminology is required to have a better understanding of the many different burn-in processes. The second step will be to examine the processes used for power supply burn-in. The actual processes used will vary by:

- Class or product (markets)
- Product volumes
- Company size
- Applications

Markets may include:

- Aerospace
- Commercial
- Computer
- Consumer
- Industrial
- Instrumentation
- Research
- Medical
- Military
- Telecommunications

Within each class, various levels must be proscribed to include the full gamut of burn-in from the minimum of no burn-in to the maximum, which may include:

- Power cycling
- Elevated temperature
- Temperature cycling
- Thermal shock
- Cold soaking
- Vibration
- Shock
- Humidity
- Radiation
- Altitude/pressure
- Powerline perturbations

The level of burn-in required may be specified in such a manner, that for some levels, in some product classes, actual burn-in may be reduced or eliminated, by a controlled process that will continue to ensure the quality of the product.

Process specifications must include criteria for the evaluation of both the actual burn-in and the process. This will aid in the observation and documentation of design and production anomalies, as well as point out concealed problems with components. In the fundamental analysis burn-in is used to effect product reliability. In too many cases, the operation is performed at typical conditions, and no real stressing of the product occurs. Let's first define the process phases for burn-in.

- Phase One: Pilot Production
- Phase Two: Production Startup
- Phase Three: Process Improvement
- Phase Four: Production Maturation
- Phase Five: Ongoing Reliability Assurance

Let's examine each of these process phases in more detail

Phase One: Pilot Production

Burn-in for pilot production must be the most strenuous phase because it is testing not only a new product, but usually a new or different process as well. Due to the low volumes, longer burn-in times are more easily achieved at this phase of the project than they would be for higher volumes.

Phase Two: Production Start-Up

At production start-up the process starts to settle out, the process and product anomalies are discovered and the root cause fixed. Sometimes additional tests or process steps are added with the goal of driving fault detection back to the point of fault generation.

Phase Three: Process Improvement

This process of improving on the process continues through ramp-up, creating a flexible mature process that produces quality products at reasonable costs. The key to success is the constant timely feedback received from the process that allows the fine-tuning of the process and the product to approach perfection.

Phase Four: Production Maturation

As the process and product mature, process stability increases, the level of attention needed to tweak the process and/or product decreases to the point where only occasional attention is needed as maturity is reached. Occasional pertabations in product quality may dictate temporary changes in the process or process parameter until the source of the problem is located and fixed. Once stability is reachieved and assured, the process will settle into the steady state mode again.

Phase Five: Ongoing Reliability Assurance

As a check of the process, some percentage of the product produced by the process is routed through additional burn-in and testing. As products and processes mature, burn-in times may be reduced or even eliminated. The Ongoing Reliability Assurance program provides a longer burn-in and perhaps additional testing to test the process.

Goals

The types of measurements performed are determined by the types of units under test, the failure modes of concern, the precision of expected results, the accuracy of measurement tools, the costs involved, the design tolerances, and importance of parameters, among other factors.

Feedback

It is possible to use the test and failure data to statistically determine the shape and length of the initial part of the bathtub curve. Some situations are extended to "life-testing," where the intent is to find the statistical life expectancy of a particular type of component or assembly, and thus the shape and starting point of the final part of the curve. In addition, burn-in may be used to accelerate early component failures, so that surviving components may be reasonably expected to be in the flat portion of their life-cycle curve. In the statistical instances, sampling is done to provide enough data for analysis. In the final case, "100% testing" is required to adequately isolate early component failures. The implementation of burn-in for statistical purposes may be a temporary process step, which is eliminated when the statistical maturity of a process and the components is assured.

Depending upon the product and the components used, and upon factors such as typical use environment and penalties of failure, burn-in may be implemented at one or more places in a design/manufacturing process. Military or aerospace applications demand that all components be in their "normal life" portion of the bathtub curb, and that they be highly predictable in terms of minimum lifetime. This regimen is necessary because military or aerospace products may be used in harsh environments, and in situations where human life depends directly upon their operation. In the case of satellites and space probes, repair cannot be done after launch, so absolute reliability is paramount. Commercial products have different burn-in requirements, because commercial or assembly is not potentially life-threatening, and most products are in situations where repair may be effected. As would be expected, new products require more burn-in for analysis than traditional products. Very mature processes often do no burn-in for analysis, reserving the burn-in process step for precipitation of early failures. In the case where all components are purchased as "pre-burned-in" by the manufacturer, the burn-in step may be eliminated entirely from the mature production process flow, when all process faults have been eliminated. Countering this trend is the desire for increased quality assurance, confirmed by continuing burn-in, which pays back in reduced failure during warranty and perhaps allows a lengthened warranty period.

The cost of burn-in stems from equipment purchases, floor space required, burn-in station maintenance, direct labor to perform burn-in potential reduction of process throughput, and skilled analysis of results. In products with reasonable volumes, these costs are generally insignificant. At low production volumes, burn-in becomes a considerable portion of production costs. The costs associated with burn-in may be plotted as an inverse exponential function with respect to volume or time. This is due to the same learning-curve effect mentioned earlier.

Burn-in is most effective when process feedback is applied. The information to correct a process or component deficiency will be best used if timely analysis is done. By proper analysis, the process immaturity factor may be controlled and the process quality will become understood and statistically predictable. Burn-in is a reliability tool, and as such, has the potential to monitor and enforce product reliability.

As for the future; power products will become inevitably more complex and greater reliability will be demanded. The tools of burn-in, as it is practiced today, will not be adequate. More degrees of variation in line quality and load excursions will have to be accommodated. The volume of multioutput power supplies will necessarily increase. The question will not be whether or not to perform burn-in, but how, when, where, and to what extent.

A technique for solving the burn-in complexity demand for power products is available but un-implemented today. Since the early 1970s, a number of companies have been manufacturing electronic loads. And since that same time, microprocessors have become inexpensive components. By using a dedicated microprocessor to monitor and control a burn-in station using electronic loads, a cost-effective means of performing complex power product burn-in may be achieved.

To perform the necessary control and monitoring, the system would have to control a number of electronic loads. These loads may be specifically designed into the burn-in controller to save space and cabling. An analog data acquisition section would facilitate measurements of the voltage and current impinging upon the loads. A method of electrically modulating the loads, to simulate typical use would be required. While adjustment of the line voltage applied to the power product is very desirable, it may be economically feasible. Controlled interruption of the line voltage is both desirable and feasible. The burn-in controller should be capable of making decisions about the health of the unit being tested, so that a shutdown sequence can be entered upon catastrophic failure. A minimum of controls are required: a start switch, an abort switch, and a display of unit-under-test status. A method of presetting the acceptable limits of measurements is required, but these parameters may be designed into the firmware of the microprocessor, or, in the case of a flexible unit, programmed from a terminal or IEEE-488 bus.

Burn-in will be a part of power product design and manufacturing processes for all of the foreseeable future. The prudent use of burn-in where practical can only serve to improve product quality and process maturity.

Life Testing and Design Verification

Life Testing The goal of life testing is to confirm the predicted longevity of a product design. The life of a product is measured by the amount of time from first application of power until a failure occurs. Life testing may be accomplished in real-time or accelerated time. Real-time testing requires that the product may be operated under some predetermined set of conditions until a statistically significant number of failures are accumulated. An average of the elapsed time each failed unit operated becomes the Mean Time to Failure.

Accelerated Life Testing   Real-time testing consumes a significant amount of actual time, and reliability engineers often do not have the luxury of time to perform real-time testing. Accelerated time testing uses stress in the form of temperature, power, vibration, or humidity to induce a failure. There are published tables that list the amount of time acceleration to attribute to each type and degree of stress applied which can be found in MIL Specification 217.

Equipment Needed   Life testing requires some method for providing the conditions in which the product will be operated, and some method for monitoring the condition of the product under test. A source of power and a device to indicate the event of failure (voltmeter, error indicator lamp) are the minimum requirements. A means of controlling the environment (temperature, humidity, and vibration) is frequently used. Power supply testing may involve loading circuitry where a shorted or open product output is not the preferred normal or stress environment.

Choosing Tests   Failures may take many forms, and there are tests to detect them by monitoring the critical specifications of the product. Life testing may be a simple periodic check of a power supply output voltage, or may be as complex as a multichannel continuous monitor of a dozen or more parameters. The definition of a particular product's acceptance criteria will determine the extent and types of testing that must be performed during life testing.

Voltage Regulation   Measurement of the output voltage is the most common measurement used in life testing. This parameter is, after all, the primary product of a power supply. The ways in which power supply outputs may be characterized are in terms of:

- Nominal output voltage
- Load regulation
- Line regulation
- Line-load regulation
- Transient regulation

Nominal Output Voltage (Single Output Supply)   The output is loaded to the maximum amount of current and the voltage is monitored to determine whether it is within specification.

Load Regulation (Single Output Supply)   The amount of current drawn from the output is adjusted from the minimum to the maximum and the voltage is measured at the extreme limits of current. The voltage at the extremes is compared to determine the amount of difference from the two current points. During this test, the voltage applied to the power supply input is held constant. The results of this test indicate how output current affects output voltage.

Line Regulation   The output load is held constant as the voltage applied to the power supply input is varied from one extreme to another. The output voltage is monitored and compared to predetermined limits. This test measures the effect of the range of input voltages upon the output voltage.

Line-Load Regulation   The current drawn from the output and the input voltage are varied from one extreme to another and the effect on the output voltage is measured. This test can indicate the ability of the power supply to operate in the worst-case scenario, which is often low input voltage and maximum current output.

Transient Regulation (single output supply)   The input voltage is changed rapidly from one voltage to another, and the output voltage is monitored for compliance to the specification. These changes may include simulated surges, voltage spikes, and missing cycles (momentary power interruptions). The transients are very short lived phenomena, lasting considerably less than 1 S each. The slew rate of the regulation control loop and the put surge protection circuitry are exercised by this test.

Multiple-Output Supplies

For multiple-output supplies, the tests become more complex. Life testing for multiple output power supplies frequently exercises the various permutations of load and source variations. A multiple-output power supply will have one output which is considered the main output, because of its dominant effect on the characteristic of the product. The main output generally will produce more power than other outputs on the same supply, or will affect the ability of the unit to regulate voltage to a greater extent than the other outputs. The main output will be tested in a manner similar to the single output supply described above.

Line Regulation The input voltage is varied, and a specified current (often the maximum allowable) is drawn from the main output. The main output is monitored to determine if the voltage is held within the specification. During this test, the other outputs from the power supply are loaded to some predetermined constant current.

Transient Regulation The input is subjected to brief sudden changes in the input voltage, as described earlier, while the current drawn from the main output is maintained at a specified level. The main output voltage is examined for deviations from the acceptance criteria. The auxiliary outputs of the power supply provide a predetermined fixed current during this test.

Load Regulation The input voltage is held constant at a prespecified value, while the current drawn from the main output is varied from the minimum allowable to the specified maximum. As before, the output voltage is monitored for compliance with the specification, and the other power supply output loads are held constant at some specified current.

Load Cross-Regulation The auxiliary outputs of the power supply are varied as the input voltage and main output loads are held constant. The effect upon the main output voltage is measured.

Auxiliary output voltages from a power supply provide less power than the main output, and are seldom used to control the regulation of an off-line switching supply. This subordinate status often results in relaxed regulation specifications for these outputs in comparison to the main output of the supply. Some power supply designs have no regulation circuitry for the auxiliary outputs, and the output voltages found there will vary with changes in line voltage, and the current drawn from the outputs of the supply including the main output. The tests performed may include (but are not limited to):

Line Regulation All outputs of the power supply are loaded to a pre-determined constant current. The input voltage to the power supply is varied over the specified operating range, and the effect upon each of the auxiliary output voltages is noted.

Transient Regulation   All outputs of the product have a specific constant current drawn from them and the output voltage from each is measured as the input is subjected to various brief and sudden voltage disturbances specified as acceptable by the product design.

Load Regulation   A specific constant voltage is applied to the input and the current drawn from each of power supply outputs, in turn, is varied, while the remaining output are held at some specific pre-agreed current. The voltage at the output whose current is being altered is measured for acceptability.

Load Cross Regulation   As with the load regulation test, the input voltage is held constant. The current drawn from one or more outputs is changed while the voltage at all outputs is monitored. This test measures the effect of current drawn from one output upon the voltage supplied by other outputs. A test of all permutations of minimum, maximum, and nominal load is often specified.

Ripple and Noise   Other than the output voltage, the amount of small signal perturbations, or ripple and noise, are commonly measured. These measurements can indicate a number of nascent failures or potential problems including loop instability, incipient failure of filter components, thermal sensitivity of one or more devices, damaged magnetic components, missing or damaged references, poor signal grounding, or unexpected AC leakage paths. Ripple is defined as disturbances of the output voltage that correspond to some periodic signal such as the input AC frequency, the regulator loop frequency or any other relatively constant signal. Noise, on the other hand, is either random in nature like Johnson (thermal) noise, "white" noise with a gaussian distribution, or bandwidth limited "pink" noise, or it may be unpredictably periodic such as unexpected resonances or external interference. Together, ripple and noise are termed periodic and random disturbances or PARD. Measurements of PARD may require a simple AC-coupled voltmeter applied to the output or may be as sophisticated as a spectrum analyzer or transient analysis system. The AC voltmeter will integrate any disturbances, and cannot be used to detect short-lived phenomena, although it is a very-low-cost monitor to use. Monitors that are able to capture and analyze the frequency, shape, and duration of brief disturbances are more expensive.

Measuring Ripple and Noise   The measurement of ripple and noise on a single-output supply involves connecting a suitable monitor device to the output. The power supply delivers its maximum rated current while powered from a lower than normal input voltage in most cases, although some power supply designs require testing at other conditions of load and input. The condition of maximum load is one which is most disadvantageous to the output filter capacitor of a power supply, and the minimum input voltage is similarly difficult for the input filter capacitor. The combination of low input voltage and high output current requirement will exacerbate any resistive effects within the regulator stages by accentuating the resulting voltage drop as the regulator must operate under dual stress. Noise tests, because of the random nature of noise, require a significant degree of noise source analysis and sometimes an understanding of how to trigger noise sources into revealing themselves.

Multiple-output supplies may require a more robust ripple and noise test strategy. The stimuli that will produce the maximum effects are as individual as each design. Careful analysis with simulation tools may help a power supply designer understand the critical parameters and combinations of input and output signals that will best initiate a ripple or noise disturbance.

Input Power Cycling   The modern power supply contains circuitry to minimize instantaneous power demand (often termed "inrush current") and AC line powered designs have circuits to adjust the periodic current demands from each cycle of AC so that the power factor is reduced to a manageable level. The operation of the inrush circuits is exercised during life testing by periodic removal and later reapplication of input power. The amount of time between such interruptions, the duration of the interruption, and the number of interruptions will depend upon the design of the inrush-limiting circuit and the specified operating environment. The input current is measured during the first second or so of operation after each application of power to determine the magnitude of the inrush current as well as to detect any long-term degradation of the inrush limiting current. Long-term failures may evidence themselves as gradual increases in the maximum inrush current from one interruption to the next. A design may require power on/off cycles as often as every few minutes to simulate the environment specified for the product, or the cycles may be as infrequent as once a day if the design is specified for a high power availability environment. The design of the inrush circuit is different for a rapidly cycling environment than that of a reliable power environment.

Peak Input Turn-On   The peak input current and power may be measured in some life test cases. This measurement allows analysis of the transient reactance effects of the power supply input in parallel with the normal input impedance. Peak input current is measured during the first full cycle after power is applied, and will capture the current required to charge an initially discharged input filter capacitor. The input power will help determine the instantaneous power factor effects during power-up.

Load Line The voltage and current may be captured as power is applied to measure the dependence of one parameter upon the other. An ideal power supply exhibits a resistive load-line, where the input current is a direct factor of the input voltage. This assumes that the power supply impedance is constant during changes of voltage and current. Most power supplies do not exhibit the resistive load-line effect over the entire range of input voltage. Measurements that fall outside the load-line specification can indicate impending regulator failure, nonlinear resistive effects (perhaps in the inrush management circuit), or damaged magnetic components.

Power Factor When the voltage and current envelopes do not coincide, the problem is described as one of power factor. Instantaneous measurement of the voltage and current on the input may be analyzed to determine the actual power versus the VA power. The VA power is the Volts Amp product. A power factor near 1.0 is the goal of any good design, designating a product that has the input voltage and current waveforms perfectly aligned as they would be on a purely resistive device. Power factor correction circuitry is available that utilizes integrated circuits (ICs) such as the ML4821 and ML4822 which are available from Micro Linear Corporation, San Jose, Ca. Power Factor Correction and Pulse Width Modulation can be had in a single chip, such as the Micro Linear Combo Controllers, including the ML4824, ML4826, & ML4841. Regulatory agencies are demanding power supplies with cleaner current draw from the line, such as better power factors. Testing for power factor may be required at some time in the future in some limited instances.

Efficiency The amount of power delivered by the output or outputs of a power supply divided by the input power is a measure of the efficiency. The remainder of the energy is lost as heat in the semiconductors or magnetic devices. High efficiency numbers close to 1.0 (100%) are the goal. The efficiency of a power supply will vary as the output current demand changes, because at low demand a large fraction of the input power is lost in magnetic leakage and transient absorption, while these losses remain relatively constant at higher currents and thus are a smaller percentage of the total. At higher currents, series resistance effects (I-R drops) become the predominant limit to efficiency. Tests of efficiency during life testing will uncover thermal damage that induces I-R drops and any damage to magnetic or transient absorption devices.

Power Device Current   Heat is the enemy of power devices. Heat is generated by the current passing through the resistance or reactance of the device. In the "off" state, the resistance is high as is the voltage across the device, but the current is near zero so no appreciable power (heat) is generated. In the "on" state, the current relatively high, but the resistance is very low as is the voltage across the device, again generating little heat. Most power is dissipated by power devices during turn-on and turn-off, when the voltage resistance and current are somewhere between their minimum and maximum values. Ideal power semiconductors require zero time between the fully nonconducting and fully conducting states, but actual devices have finite and often lengthy turn-on and turn-off times. Resistance falls, current increases, and voltage decreases as the device turns on. The voltage rises and the current decreases through a power device as its resistance increases during turn-off. A decent life-testing strategy includes the measurement of power device current during turn-on and turn-off, as well as a measurement of the time required for each edge of a typical cycle. Some power devices become slower as they heat up, and thus spend a larger percentage of each cycle in the high-stress region of the edges. This test can detect increased resistance by detecting lower currents as well as 'thermal runaway.'

Over-Current Measurements   A power supply is designed to produce a certain amount of output current from each output. A power supply may be damaged if it is required to produce significantly larger current than the design limit. One mechanism for damage is the heat generated as a result of the current passing through the resistance of the regulator circuit. Many power supplies incorporate circuitry to protect the regulator when too much current is requested. Life testing often verifies the continued operation of these protection circuits.

Foldback Outputs   One protection strategy reduces the output voltage below the normal value. A fixed very low resistance load is assumed. Reduced voltage will similarly reduce the current through a fixed load resistance. The output voltage remains below normal even after the excessive demand is removed. The name for this strategy is foldback current limiting. This regulator protection technique significantly reduces the amount of power and heat generated when the foldback circuitry is activated as compared to the normal operation. The potential for thermal damage is decreased once the foldback circuit is triggered.

Overcurrent Threshold   The power supply designer must select a specific value of current for the foldback circuit to be triggered. Current beyond the threshold setting will cause the output voltage to drop below normal.

Elevated Temperature Overcurrent
Heat is a primary mechanism for regulator damage, so operation of the product at elevated ambient temperature leaves a smaller margin for heat generated by excess current. Some power supply designs compensate the current limit threshold as the ambient temperature increases. Many newer power ICs have a thermal shutdown feature to prevent device damage. Compensation circuits tests over a range of temperatures are often performed during life tests.

Prolonged Short Circuit
Continued operation in short circuit or overcurrent foldback mode may require testing. Some designs cannot tolerate indefinite operation in overcurrent foldback.

Recovery from Overcurrent
The overcurrent test must remove the input power source for the power supply to reset the overcurrent foldback circuit. Foldback designs will remain in reduced voltage mode until the input power has been interrupted and restored. A recovery test is performed to verify that the current foldback was not the result of regulator damage.

Current-Limited Outputs
Some power supplies use a strategy that differs from the foldback technique of protecting the regulator. A current limited output changes from constant voltage below the set-point to constant current output beyond that level.

Overcurrent Set-Point
The voltage on an output will remain constant with changes in current until the overcurrent set-point is reached. Current demands beyond the set-point will cause corresponding reductions in the output voltage. The output voltage will return to normal as the current demand returns to the acceptable range below the set-point.

Elevated Temperature Set-Point
The overcurrent set-point is lower at an elevated ambient temperature on many supplies that it is at room temperature. This change in set-point compensates for the inability of higher temperature air to dissipate higher temperatures from inside power devices.

Prolonged Short Circuit
The voltage on a current limited output will drop to near zero as a short circuit is applied. The power dissipated by

the regulator will be near maximum during a short circuit, as the voltage difference across the regulator is the full value of the input voltage and (by definition) the maximum current that the regulator may provide. The temperature of a shorted regulator will reach its highest value during prolonged operation.

Recovery from Short Circuit   Current limited outputs will recovery their normal output voltage as the current demand is reduced to within the nominal output current range. A life test may verify that the voltage returns to normal after removal of the overcurrent condition. A product that fails to respond to removal of an excessive load may indicate damage to the regulator or the overcurrent protection circuit.

Pulsed Overcurrent Outputs   The foldback circuit described earlier reduces the thermal energy dissipated in the regulator during excess current demand, but requires a removal of the input power before the output voltage is allowed to return to normal. The current limited output does not require a power off cycle to reset the protection circuitry, but the thermal result of current limiting is much higher heat generation during current limit. Another method for protecting power supply regulators is the pulse overcurrent technique. Here we have two levels of overcurrent protection. The first is the normal overcurrent type of circuitry. The second allows a higher overcurrent for just a short period of time. This type is used for motor and other similar loads where starting current can pulse above the normal maximum current ratings for every short periods of time. After a short delay of a few milliseconds, the higher overcurrent level is deactivated and then the normal overcurrent circuitry is allowed to continue operation. The thermal energy dissipated in the regulator is lower than normally would occur if the higher value of current limit were used steady state, thereby reducing the potential for component damage.

Peak Output Current Measurement   The pulse method reduces power dissipation by significantly reducing the average current after the demand triggers the protection circuit. When protection is engaged, the peak current value of each pulse continues to increase as the overall average current sharply decreases. The eventual maximum current available during each pulse is limited only by the internal series resistance of the regulator. The measurement of the peak instantaneous current during short circuit conditions will demonstrate the equivalent series resistance of the regulator circuit.

Average Output Current Measurement   The average value of output current will increase as a function of increased demand until the overcurrent set-point is reached. After the protection circuit is activated, the average value of current will decrease by a large margin. Measurements of the average current during short circuit indicate the amount of pulse energy that will generate heat after protection circuits activate.

Overcurrent Set-Point   The overcurrent set-point may be measured by slowly ramping the current demand from nominal to excessive values and monitoring the output current averaged over periods of time about 50 ms or less in duration each. The maximum current value detected during ramp-up will indicate the overcurrent set-point.

Elevated Temperature Overcurrent   The regulator will generate more average thermal energy (heat) at the current level just before the overcurrent threshold is reached. Increased ambient temperature may reduce the ability of the thermal management systems (heatsinks, fans, coldplates) to remove sufficient heat to keep the power devices within their design specifications. The overcurrent threshold may be designed for a negative temperature coefficient; that is, it may lower the overcurrent set-point as the ambient temperature rises. A life test system will confirm that this temperature dependency operates correctly.

Prolonged Short Circuit   Correctly designed pulsed overcurrent outputs are relatively insensitive to prolonged short circuits at their outputs. The mechanism for regulator damage from prolonged short circuit is the peak current borne by the power devices. Repeated current peaks near the design limit of power semiconductors can damage the devices by inducing microscopically localized thermal effects, which will eventually produce electrical failure before the case temperature of the devices reach a critical point.

Recovery from Overcurrent   Most systems are designed to resume normal operation as the current falls below the overcurrent set-point. Some designs may introduce some measure of hysteresis, so that the overcurrent protection circuit activates at one current, but will not deactivate until the demand falls well below the activation set-point current. The set-point for most designs is well above the normal operating specification, so a current demand within the specification is usually below the hysteresis recovery point.

Dropout Capabilities Power supplies do not generate power, but simply manage the conversion of that power from one level or form to another. The power input may be unreliable or unpredictable, and in those cases the power supply reaction to losses or reductions in the source of energy may be important. There is a point below which a power supply cannot produce an output within its specification due to insufficient input power. Life testing may be required to test this because the effect on a product operation due to electrolytic capacitor aging caused by electrolyte dry-out.

AC Input Supply

Dropout Point The nominal range of AC input power is ± 20%. Power supplies are designed to operate over that range, and life testing will often include a test where the input voltage is lowered more than 20% to determine where the power supply cannot produce an output voltage within normal specifications. During this test, the output or outputs are loaded so that they produce maximum current. The failure limit of the power supply at low input AC voltage is design-specific, but seldom exceeds 25% lower than normal input voltage. This voltage is measured as the RMS value of the input waveform rather than the instantaneous DC voltage of the AC cycle.

Pull in Point A modern power supply requires power for internal circuits that may include comparators, amplifiers, timers, and oscillators among other circuits. A crude internal power supply operates these circuits which in turn coordinate the signals on the product itself. The internal power supply is designed to operate over a very wide range of input voltages, so that it may operate at voltages too low for the normal product to operate and will thus provide power to internal circuits before the input power has reached the level where full produce operation is possible. The life test may confirm that the input power can be ramped up to a point where the internal power source begins operation but the product is not operational, and then a small increase in input AC voltage enables the product itself.

DC Input Supply

Dropout Point Direct current input supplies are designed for a wide range of environments. Power supplies designed for rectified AC operation must operate over a range of inputs that will result from the rectified and filtered AC input signal. Other supplies designed for operation from batteries do not require a wide range of input voltage, as batteries have very narrow ranges of acceptable outputs. Direct current power sources will exhibit reductions in power and outright failure as well as do their AC cousins. The DC Input supply will have some specified DC voltage below which it cannot operate. Life testing will often test this by lowering the applied DC voltage to determine that the product continues to operate above this specified limit.

Pull-In Point   The life test may ramp the input DC voltage up to determine at what DC input voltage the output from the supply will meet specifications. This point must be below the normal operating specification, but is generally slightly more DC voltage than the drop out point.

Ride-Through   AC powered supplies may experience complete loss of input power for part of one AC cycle or several cycles in a row. Many designs have the ability to continue operation during the short-term loss of AC input power. The AC input power loss is defined as any point where the input voltage drops below the drop-out RMS voltage (see above). The ability of a product to ignore brief interruptions in the input power is based upon the reserve power storage of the AC input filter. The input filter attempts to provide the power needed by the power supply circuitry during periods when no energy is applied to the filter. The reserve is based on the product of energy required and time duration; smaller power demands can be met for longer time durations, and vice versa. The life test may load a power supply to its maximum current output and then interrupt the input power for half of one cycle to determine if the output remains within the specification. This test is often performed using 50 Hz AC input power (if the power supply may be safely operated at 50 Hz), because a half-cycle of 50 Hz is 20% longer in duration than a half cycle of 60 Hz power, and the longer time duration becomes the critical specification. If a product is not capable of operation from 50 Hz, or the life test setup cannot generate 50 Hz AC power, an alternate method of complete power removal may be used with a timer monitoring how long the power supply output remains within the specification. The alternate method requires some analysis of the results against the design, as the output specification may be maintained by regulator output capacitance and not be a direct function of the input filter design. The ride-through test is designed to evaluate the continued health of the AC input filter circuit.

Component Stress Verification   During life testing, additional information may be collected that is not part of the published product specification but that will help the designer characterize and correct the design.

Semiconductors

Maximum Voltage Applied Semiconductors are designed to operate with voltages lower than specified limits applied to their terminals. Voltages above the specified limits may damage the semiconductor, reducing the life of the device and eventually precipitating product failure. The voltage waveform at all terminals of semiconductors including power devices and control circuits should be observed by oscilloscope or other waveform analysis system so that any transient voltages above the specification may be detected.

Maximum Current Applied The current passing through a semiconductor may be monitored at each terminal to determine if the amount of current exceeds the rating for the part. This current must be observed with an oscilloscope or waveform analysis device to capture short-lived high current events.

I^2T Measurements Every semiconductor package is designed to release a certain amount of heat. The square of the current multiplied by the time gives an indication of the amount of thermal energy that will be generated by the device inside the semiconductor package. The amount of heat must not exceed the thermal limits of the package or damage to the semiconductor will occur. The current waveform and time measurements collected from waveform analysis or an oscilloscope will allow this parameter to be measured.

Input-Induced Transients Short lived voltage perturbations on the input circuit of a power supply may find their way into critical circuits. These signals may be generated by the test setup and their effect on internal signals in the power supply noted.

Rise and Fall Times (Power Devices) The dissipation of a power device is very low when it is off because of low current, and likewise the power is low when the device is fully conducting because of low voltage drop across the device. During the transition time between off and on, and between on and off, the instantaneous power dissipated by the device climbs sharply. Power supply designers attempt to use devices which make the change from off to on and on to off very abruptly, but even the best devices have finite transition times. A life test may measure the rise and fall times a critical power semiconductors to ensure that the devices are operating as specified. Increased transition times seen at life testing may indicate incipient device failure. Causes of lengthened transition times include highly capacitive interface circuitry and poor thermal management.

Startup at Low Line Some designs exhibit difficulty beginning operation at low input voltage conditions. These designs operate normally when the source of energy is brought up to normal input voltage and slowly reduced to low line levels. A test of the ability to start operation at the lower limit of input voltage and full load will indicate whether this problem exists in the design.

Temperature Rise The efficiency of the power supply will determine the amount of heat that will be presented to the thermal management components of the product. Unexpected temperature rise at the case of critical components and on the product enclosure will indicate problems with thermal management.

Capacitors

Maximum Voltage Applied Voltage impinging upon power supply input and output filter capacitors may have a much higher peak value than the average DC voltage across than would indicate. Equivalent series resistances and inductance on capacitors may allow the voltage at the terminals of the component to reach many times the average value. Short-duration voltage excursions beyond the design limit of a capacitor will damage the device dielectric. The voltage across the input and output filter capacitors could be measured with an oscilloscope or waveform analyzer to confirm that the voltage rating is not being exceeded.

Maximum RMS Current Applied The equivalent series resistance of a capacitor will translate a portion of the charge or discharge current into heat. This heat is detrimental to the continued operation of the capacitor. Electrolytic capacitors will evaporate their dielectric electrolyte in the presence of heat. The current seen by capacitors, particularly electrolytic devices, during discharge and charge should be quantified.

Temperature Rise As stated above, heat damages capacitors. An elevated temperature on a capacitor may indicate a design problem or a defective part. Capacitors may be subject to large temperature rise from external heat sources like physically nearby power semiconductors or heatsinks. External heat will damage a capacitor as surely as an internal temperature source. The temperature of all electrolytic capacitors should be monitored to ensure that they do not exceed the specified operating range specified for the device.

Resistors

Maximum Wattage Continuous   The continuous power rating for resistors is based upon the ability of the device to shed heat. The maximum average voltage and current for resistors should be measured to compute the maximum wattage the device must dissipate.

Inductance   Resistors that are spiral cut carbon or metal film construction will exhibit some inductance. In some circuits, such as switching regulators, these inductances can add problems to the circuit, especially if the inductance is located where it doesn't belong. Examples include: FET gate drive resistors and current sense resistors, just to name two areas where problems can occur.

Maximum Peak Power   Resistors have some finite thermal mass, and are able to absorb a limited amount of short-term energy above their continuous power rating. Energy of sufficient magnitude or duration will damage the device due to the converse of thermal mass, which is thermal resistance. The thermal resistance from the location of thermal stress to the lower-temperature thermal mass must be small enough to maintain the temperature at the stress point below some predefined level. Many resistor manufacturers specify the maximum peak power that their resistors will absorb (and for what duration). A design that stresses resistors with brief pulses of power above the continuous power rating for the resistor must be tested for compliance to the resistor manufacturer's specification.

Transformers

Maximum *ET* Product   ET is the volt time product of pulse transformers such as used in FET gate drives. The Time T is usually measured in microseconds. Magnetic flux in transformer cores will generate heat. The energy impinging on the core may be measured and the efficiency of the transformer determined. The loss of efficiency will translate directly into heat.

Maximum Temperature Rise   Properly designed transformers are very efficient and produce little heat. Excessive temperature rise on a mag-

netic component such as a transformer may indicate poor design or incipient failure. Heat may be created in transformers by unexpected direct current that elevates the temperature of the transformer windings or extremely high-frequency energy within signals that contribute to magnetic core losses. Either DC or high frequency energy problems are serious design flaws that should be thoroughly investigated when detected.

Special Feature Testing

Power supplies may generate status signals or other electrical information. Some power supplies accept control signals from external circuitry, and a few products require external electrical stimuli other than the power source for proper operation.

Monitoring Signals The auxiliary signals should be monitored or supplied as needed for the product in question. Some of the auxiliary signals found on power supplies include:

- AC Low
- DC Low
- Buss Low
- Temperature Outputs
- External Clock Input
- External Clock Input
- Voltage Reference Input
- Voltage Reference Output
- Current Mode Control Voltage Input
- Current Mode Control Voltage Output
- Overtemperature

AC Low activation indicates when the input AC power is too low to sustain any product operation. This signal may be derived directly from the internal power source for the product or may be simply connected to a comparator which monitors the voltage on the input filter. AC Low is commonly designed with hysteresis so that the RMS value of the AC signal must return to a value well above the value where the "too low" signal occurs. This signal is commonly found on power supplies that operate a device that performs some electrical sequence upon power application and removal.

The **DC Low** signal becomes active when the internal power supply circuits have detected marginal conditions for operation. DC Low usually activates a few milliseconds before AC low. Upon power application, DC Low deactivation follows the deactivation of AC Low by a similar delay. DC Low activation usually serves as an early warning of power loss. DC Low is usually derived from the same circuit as AC Low, but with a different threshold value.

Bus Low is a signal found on a computer power supply, which indicates when the output power can be maintained only within a limited range of loads. This may be used in conjunction with AC Low and DC Low to perform an orderly shutdown of machine functions in the event of power failure. Like AC Low and DC Low, this signal is derived from a comparator, and may be part of the AC Low and DC Low circuit.

Temperature output is usually an analog signal that reflects the temperature value somewhere in the power supply enclosure. Complex systems monitor the power supply temperature and shut down sections of the load or the supply when the temperature gets too high. The placement of the temperature sensor and the description of the output signal are both design-dependent.

External Clock Input is used on multiple-regulator supplies that are synchronized to some reference frequency. Standalone power supplies seldom include this signal, but custom regulator modules may use it. External clock output is provided from the so-called master module, and this signal is used as the input to all of the slave units in multiregulator switcher designs. This is sometimes done to control EMI/RFI.

A **Current Share Bus** may be present in multimodule power supplies, or they may share one voltage reference among the various regulator modules in the system. One module in the system may become the master reference source, and all of the other modules are slaves to that signal. On some systems, there is no master designated, and the unit with the highest output voltage becomes the master unit. Each module in the system will then equally provide the required power output.

Overtemperature signals are provided as a warning to an operator control panel or a monitor circuit external to the power supply. These signals are generally logic signals and do not indicate the actual temperature but merely the acceptability or unacceptability of the temperature at some point within the power supply enclosure.

Control Signals are among a number of other signals that are found on power supplies to control the operation of the unit. These signals may include:

- Remote Enable
- Remote Sense

- Remote Voltage Adjust
- External Current Sense
- Overvoltage Protection Indicator
- Over Current Indicator
- Ground Current Monitor
- Overtemperature Monitor
- Air Flow Sense
- Fan Power
- Battery Backup Signals

Air Flow Sense is usually accomplished by using either a vane switch in the air flow path or a thermistor or other temperature sensing element in the air path. The purpose is to prevent thermal damage when filters become clogged or when air flow stops. In some systems air flow is provided to the power supply from an external source. The power supply can be designed to either not operate without air flow, or to operate at reduced rates.

Remote Enable is a logic signal input or switch closure input that allows the user to turn the power supply output on and off without interrupting the input power source. Remote enable is commonly used during extended tests to provide controlled shutdown of a filing power supply.

Remote Sense is an analog input that connects to the point for use for the power from the power supply regulator. The remote sense voltage is monitored by the power supply instead of the voltage at the power supply output terminals in an attempt to compensate for resistive voltage losses in the cables between the power supply and the point of power use. Extended testing should verify that the voltage delivered to the point of use is relatively insensitive to the resistance of the power cabling.

Remote Voltage Adjust is an input found on some power supplies that must be adjusted after assembly. This input may also be used when the adjustment would involve access to enclosed components and controls. These signals should be tested for the ability to adjust the output voltage over the specified control range.

Many power supplies use some form of **Overvoltage Protection** at their output to protect the load in the event that the regulator fails by shorting the input voltage directly to the output. These overvoltage currents are designed to rapidly shunt the output to ground. In many instances the overcurrent circuits activate and shut the regulator off, but when these circuits cannot operate due to complete regulator failure, the power supply input protection (fuse or circuit breaker) may activate. The Overvoltage Protection indicator is used to signal when the overvoltage protection circuit is activated. Many systems monitor this output of the power supply to alert a user or other alarm hardware. This signal should be tested to ensure that overvoltage events will be reported.

Overcurrent and **Overvoltage** indicators and/or monitors are found in some power supplies. A monitor signal or indicator device indicates when an overcurrent or overvoltage condition has occurred.

Ground Current Testing may be required for some applications, to measure the current flowing between the power supply and the connection to ground. Appreciable ground current indicates serious problems, and must not be ignored. For patient-connected medical equipment and systems, power supplies are designed to meet the leakage requirements of UL-544.

Overtemperature Monitoring may be required for thermal management of many products, that depend upon sufficient airflow through the power supply enclosure for proper operation.. Designs with a critical airflow requirement through the power supply enclosure may require the use of an external airflow sensor to indicate when air ducts are blocked or a fan has failed. Signals indicating reduced airflow should initiate shutdown of the power supply. Some power supplies may operate external systems with critical airflow requirements, and in this case they will supply information about the airflow within the power supply to those external systems so that they may decide if sufficient airflow exists.

Fan Power is provided by some modern power supplies which may require extensive cooling, even during battery backup operation. Additionally, some fans may be very noisy. One method of quieting the mechanical noise of fans is to operate them with DC power. Some systems operate the fans with AC other than the 50 or 60 Hz line power. The obvious source of the power for cooling fans is the power supply. Testing should be included to confirm the voltage and available current for the fans, and that any AC signals are the proper shape and frequency.

Battery BackUp Signals are becoming more common is complex power systems. A power supply design may incorporate circuitry to allow operation with batteries. In addition to the regulation and conversion of battery power, some power supplies are able to recharge the batteries and provide status signals to a control panel or monitor circuit. A robust test suite will include tests of the status and control signals under discharge and charge conditions as well as with a variety of levels of battery charge.

Adjustments may be required on some power supplies; therefore, a power supply may contain one or more adjustable components. Potentiometers are commonly used to control the voltage setpoint, overcurrent trigger set-point, overvoltage trigger level, or any of a number of other parameters that are defined by the user or cannot be controlled with enough precision in a generic design. Tests that measure the range of adjustment and the final setpoint of these adjustable components are an important part of a power supply test.

Mechanical Testing is required for some power supply subsystem units to verify the overall size or mounting holes or hardware locations. In these cases it is not uncommon to use a mechanical test jig or fixture to facilitate the measurement process.

Safety Testing is a common test of power supplies that measures the quality of the insulation between the power leads and the enclosure (ground). This test is termed HIPOT testing, and is conducted by placing a high voltage across the ground and the power leads and then measuring the amount of current that passes between the two elements. This tests the amount of leakage current allowed by the particular international standard of interest. For some applications insulation may be measured in megohms at some DC voltage (usually 500 VDC).

Interpreting the Data is the most important step. The results of a few tests are qualitative, and may be understood as pass/fail tests. Some tests require a numeric analysis of the results based upon the design of the product or the intended application. A failure of one test may indicate a design problem, while other tests may indicate that a component or circuit topology is inappropriate for the intended purpose. The analysis of test results will require the participation of the product designer so that design corrections may be incorporated without compromising the intentions of the product design.

5

Pre-Production Testing

Testing during a pilot production run is just as important as it is during the manufacturing process, but for different reasons. During pilot production, testing is used to isolate problems in the design that were not detected during simulation and breadboarding, as well as problems in the manufacturing process itself.

Pilot Parts Specifications

There is a natural tendency, where part selection during simulation and prototyping is strongly influenced by electrical characteristics and properties. The critical specifications at this phase involve the suitability of each part to a particular manufacturing process. Later in the process, design and selection or reselection of parts based upon their suitability for the manufacturing process is termed "Design for Manufacturability."

Design for Manufacturability

Physical shape and size are the most obvious parameters that fall under these considerations. In the design environment, physical size and shape of components are less an issue as the designer concentrates upon electrical characteristics and performance. The prototyping phase may be biased toward selecting parts which are readily available to meet schedules, although some of the parts used in a prototype run may be replaced in pilot production by electrically similar parts with different physical appearances due to availability. Thus the pilot production phase becomes the proving ground for the physical part of the design. Some of the areas of investigation are as follows.

1. Determinations are made about the physical positioning of adjustable components that require some level of physical access. Unless care is exercised, components requiring access such as dual in-line plastic (DIP) switches and trimmer potentiometers can be misplaced on a printed circuit board (PCB) design. The metalwork to PCB alignment must be properly designed into the system, or else errors may be discovered during pilot assembly. This misalignment of part and access may require redesign when discovered.

2. Subassembly to assembly alignments are verified during pilot manufacturing runs. When two or more assemblies share a common physical component such as a heatsink or connector, conditions may exist where these assemblies will not fit together as intended. Cumulative mechanical tolerances often are the culprit here, although errors in physical models in the physical design tools can occur.

3. Physical design errors such as bulky components with insufficient mechanical support may be uncovered during pilot production. It is here that added support or additional stiffening or support elements are added as required.

4. Components that interfere with or prevent other components from being properly assembled may not be noticed until the pilot production run.

5. Where automatic parts placement is used, the sequencing of insertion (or onsertion) operations is tested to ensure that parts that may be physically obscured by other parts are placed before the parts that would obscure their later placement.

6. Component selection based on the manufacturing processes that are to be used for the product, to reduce the cost of manufacturing for the product. A part that costs a few cents more than another, but requires minutes less assembly (direct labor) time is not only a cost justifiable part selection, it is a desirable one.

Designers normally find one or more electrical problems in their design during pilot production, regardless of the quality of their design efforts. Simulation should have reduced the quantity of these design problems to zero, but even the best simulators have their limits. The types of electrical problems found during pilot manufacturing may include, but are not limited to the following.

- **Part-to-part crosstalk** Virtually no simulator available on this planet today is available to reliably predict the existence or severity of crosstalk between one component and another. Some layout simulators are able to make reasonable estimates of trace-to-trace crosstalk and these types of simulations will become more commonplace in the future. This type of design problem can be found in low-amplitude circuits adjacent to high-speed and high-energy circuits. Induced crosstalk effects may not be noticed until the point of the startup of manufacturing a new product.

- **Interpart thermal effects** Once again, simulators cannot yet reliably predict thermal effects of one part upon another. This is due, in part, to a problem with SPICE. Newer versions are expected to solve this problem soon. Component thermal output is poorly specified by almost all manufacturers, although thermal sensitivity is usually well understood and specified. A part's thermal performance may have been exemplary during simulation, but in the proximity of a part acting as a heat source, the performance of the part may degrade beyond acceptable limits. Another temperature-related error is created when a temperature compensation component is exposed to a different thermal profile than was intended.

- **Electrical ground effects** The zero voltage reference is subject to a number of disturbances due to poor distribution and electrical current fed to the reference. Since few of the electrical characteristics of reference planes are simulated well, unintentional "ground loops" will escape detection in a simulator and prototype, only to surface in the units in pilot production.

- **Instability effects** Most simulators do a very good job at predicting instability when the models are accurate enough, but these accurate models of components may not be available when needed. New or custom components can be poorly understood and models of these components can be poorly understood, and models of these components can be inadequate, if they even exist at all. Compensation and feedback paths are common locations of components that influence instability, and complex parasitic feedback paths are difficult to model accurately if they are not well understood and well documented

When "designing for manufacturability," specification details become significant. A designer tends to look for components and assemblies that are either already familiarly or exhaustively specified to fulfill the requirement. Manufacturing processes are usually somewhat intolerant to surprises that manifest themselves after the unit leaves the prototype stage and enters production. Complete specifications are a requirement to allow the designer to predict and avoid these surprises.

Once appropriate parts and acceptable alternate parts have been identified for the pilot run, this information is ideally fed back to the design process as criteria to aid designers in future designs. This is an area where many companies could do better communicating back to engineering.

Pilot Production Run Process

Pilot production runs vary in scope as widely as the final production runs. Pilot production may encompass the manufacture of one or two units when a low- or medium-volume manufacturing process is being developed, or as many as several hundred units on a very-high-volume product. Pilot production is used as a dress rehearsal for full production of a product, so the number of units produced in a pilot run is considerably fewer than will be produced in the actual manufacturing process. The watchwords for the pilot process are "control" and "feedback." Control is maintained by the designer or production engineer to ensure that the process variables being measured are not affected by unknowns in the process. During a pilot run, the production engineer attempts to identify as many production problems as possible and eliminate the problems before a unit is released to full production, where solutions are too costly. Because control is so important, the importance of testing requires a clear and comfortable understanding of test processes. To eliminate as many variables from the test portions of a process, a production engineer strives for complete and comprehensive tests. Complete testing of an assembly is accomplished when the presence or absence of every component can be detected.

Comprehensive testing implies that the components and assemblies being tested do not have critical functions or parameters that are unable to be tested. The types of tests and parameters measured by in-circuit testers are commonly considered to be comprehensive. A production engineer makes the evaluation of which parameters and device functions are critical to a circuit or design, based upon the design, experience, and perhaps some intuition. The goal will be to isolate and identify those parameters that will complicate a volume manufacturing process and nullify their effect before full production begins. A test engineer is given the test specification, and from it determines the required test hardware and software. Inevitably, a few of the components will be 'untestable', meaning that one or more of their parameters cannot be measured once they are in the circuit. It is the job of the test engineer to create tests for these so-called untestable parts. See Loop Testing in Chapter 8.

Preventing components from becoming untestable is also part of design for testability. There are several important points to remember when designing for testability:

- Avoid paralleling more than two or three components to achieve a particular value. This is particularly important when high-energy circuits are involved because of the low impedance and resistance values involved. Paralleling resistors to share current and thus withstand higher power is a common design technique used to reduce costs. As the number of parallel components increases, the level of testability decreases. Should one or more of these parallel resistors be a wrong value, the majority of current will flow into (and majority of power will be dissipated by) the lower valued resistor(s).

- Most in-circuit testers will be unable to detect the error in values, and even if they could, cannot determine which resistor in a parallel circuit is in error. Since the resistors will not be stressed during in-circuit test, it is likely that the catastrophic failure of the overstressed resistor(s) will occur when the unit has full power applied after assembly is complete.

- Avoid using components whose parameters are difficult to measure, whenever possible. For example, there are inductors that can tolerate so little current through their windings that many general purpose ohmmeters destroy the inductor when any attempt is made to measure their winding resistance.

- Where feasible, use built-in test aids. Recent advances in custom and semi-custom integrated circuits allow these circuits to either perform tests on themselves or to give external test systems a "window" into the circuitry deep inside these components. A tester that is able to stimulate and measure nodes inside these circuits is able to more completely examine the circuits for proper operation. Although these built-in test capabilities are usually associated with logic circuitry, increasingly analog and mixed design integrated circuits are using built-in test.

Pilot production is a shakedown for the test processes as well as the assembly processes. A test engineer often uses pilot production to characterize the instruments, procedures and interfaces used for test particularly when unfamiliar components or unusual test techniques will be used. A designer wishing to facilitate pilot production will work closely with the test engineer throughout the design process to ensure that effective tests can be developed in time for pilot production.

Pilot Parts Qualification Testing

One key to successful pilot and volume manufacturing is designing for quality. Quality is an elusive term, best redefined as "predictability." A predictable process or part is one that is well understood, understanding coming from either documentation or experience.

Quality is often associated with increased cost or a longer time investment, but high quality can be achieved without incurring either penalty. The cost of correcting an problem in a unit rises dramatically as the problem is discovered later in a product life cycle. An exponential relationship between where in a process a problem is discovered later in a product life cycle. An exponential relationship between where in a process a problem is discovered and the cost to correct the problem is commonly accepted. In Figure 5-1 this relationship is depicted. A problem that would cost $1 to correct in the design phase would cost $10 to fix in the prototype phase, $100 in the pilot production phase, $1,000 in volume production, and $10,000 to correct once shipped to a customer. Experience shows this curve to be relatively accurate as a measure of relative cost. It can be seen that a problem that takes as little as 4 hr to fix in the design phase could escalate into a problem costing millions of dollars if it cannot be discovered before customers receive the product. The cost of the product during its life cycle is a primary component of the price of the part, and in a free-market economy, the manufacturing costs determine the amount of profit that a particular part will generate. Experienced designers strive to detect problems as early in the life cycle as possible to maximize profit, without seriously affecting time to market.

Information fed back to earlier phases, and actions taken to correct problems contribute mightily to the steepness of the cost curve. Attempts to avoid added cost by subverting the feedback process lead to correction costs that may exceed the penalties associated with feedback and these attempts are thus ineffective at cost containment. It is a rare manufacturing situation where this curve does not apply. Designers ignore this fact at their own risk.

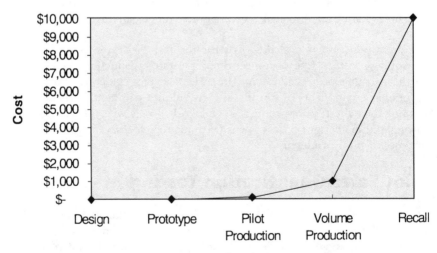

Figure 5-1.   The cost of defects to a project.

If designers were able to perform their task perfectly, there would be no need for prototyping or pilot production. In the real world these steps serve to assist the problem detection effort.

The pilot production phase is the watershed point in the life cycle, where problems may still be discovered without serious adverse effects upon the cost of the product. Beyond the pilot production phase, any problems will increase costs associated with inventory and product rework.

Design for Quality

There are several strategic for achieving and maintaining high quality in a production environment, and the conscientious application of these strategies will easily repay their costs in reduction rework and inventory reduction. These strategies are put in place during pilot production and they continue to operate throughout the manufacturing lifetime of the product.

Ship to Stock

One of the more increasingly popular strategies for improving quality is to involve part suppliers in a "ship-to-stock" program. In this technique, the testing and inspection of each part is done by the supplier under very stringent guidelines agreed upon in advance by both the vendor and customer. By virtue of production volume, the supplier is better suited to perform part qualification tests on a part than is any single customer of that supplier. This strategy often produces the lowest total costs when the costs of incoming inspection are included in the total cost of a part. You can then determine the amount of profit that a particular part will generate. Experienced designers strive to detect problems as early in the life cycle as possible to maximize profit, without seriously affecting time-to-market.

Information fed back to earlier phases, and actions taken to correct problems contribute mightily to the steepness of the cost curve. Attempts to avoid added cost by subverting the feedback process lead to correction costs that may exceed the penalties associated with feedback and these attempts are thus ineffective at cost containment. It is a rare manufacturing situation where this curve does not apply. Designers ignore this fact at their own risk. Were designers able to perform their task perfectly, there would be no need for prototyping or pilot production, but in the real world these steps serve to assist the problem detection effort.

The pilot production phase is the watershed point in the life cycle, where problems may still be discovered without serious adverse effect upon the cost of the product. Beyond the pilot production phase, any problems will increase costs associated with inventory and product rework.

In a successful ship-to-stock relationship, the customer sets quality standards and works with the supplier to decide upon a testing and inspection regimen to achieve that level of quality. The supplier implements the regimen and is able to deliver fully tested and qualified parts to the customer. The supplier gains an additional benefit by pulling defect detection and correction back into the manufacturing plant from the field, thus reducing the life-cycle costs of the part and increasing profitability. The customer reduces costs by eliminating or drastically reducing incoming inspection, and reduces costs associated with inventory because fewer bad parts will be found in inventory.

Ship-to-stock programs are very effective but there are situations where they are clearly inappropriate: when parts are purchased from multiple suppliers it becomes difficult to maintain consistency of the testing and inspection procedure from supplier to supplier. Parts purchased from distributors are often not part of a ship-to-stock program unless the distributor participates in the testing and inspection of parts from various suppliers.

Some parts suppliers publish the details of their testing and inspection processes so that customers wishing to cooperate in a ship-to-stock relationship need not develop criteria for testing parts and may simply agree to the procedures published by the supplier. This allows the supplier to develop a single qualification and testing process, and further unburdens the customer if the processes used by the supplier include all of the testing criteria that the customer considers important. The U.S. Department of Defense publishes standards[11] for the manufacture and testing of electronic components and manufacturers who make parts for military projects tailor their process qualifications upon these standards. Few commercial products demand the ability to withstand the wide range of temperature, humidity, vibration, and electrical stress that military products are required to tolerate, thus the governmental standards generally exceed the requirements specified by other customers.

Trust is a key element of any ship-to-stock agreement. Customers must have confidence that a supplier is safeguarding the quality of the parts being delivered. Since a ship-to-stock program reduces incoming inspection test costs while the cost of each part rises slightly, it makes poor economic sense to perform full inspection on parts that have ostensibly undergone identical testing at the manufacturing plant. Prudent customers periodically inspect a quantity of parts involved in ship-to-stock to reassure themselves that the supplier is indeed being honest.

Even honest suppliers can have testing and inspection process failures that result in the shipment of untested or partially tested components. A

---

[11] See Annotated Bibliography at end of chapter.

consistent sample testing process performed by the customer often will pinpoint failures of the supplier's process and provide notification to the supplier. In these instances, the retesting or replacement costs are customarily borne by the supplier, although the responsibility for process failures is an element of any ship-to-stock program contract. International diplomacy has coined a phrase for the cautious acceptance of another's word: "Trust, but verify." This phrase is no less true in the manufacturing milieu.

Incoming Inspection

When a ship-to-stock program is not possible, or where an existing supplier has demonstrated lower than acceptable quality, a customer must maintain an incoming inspection program, incoming inspection usually is characterized by one or two forms: sampled inspection or 100% inspection.

In sampled inspection, a small percentage of the received parts are tested, and statistical methods are used to extrapolate the results of the inspection to determine the characteristics of the entire lot of received parts. This technique is particularly useful where the number of parameters of interest is small and the lot sizes are relatively large. As the number of parameters increases, the costs associated with incoming inspection increase. As the number of parts in each lot increases, the likelihood that a grouping of parts conforms to a statistical model increase.

Another form of sampled inspection tests all parts in a lot for conformance to a subset of the entire specification. By definition, the number of parameters of interest is small, and since all parts in a lot are tested, the size of the lot is not important to the statistical characterization. This type of testing is often used when a small number of parameters are known to vary outside of the specification, or where a defect that is simple to detect during incoming inspection is difficult to detect until very late in the manufacturing process.

In 100% inspection, every part is inspected and tested for conformance to the entire specification. Obviously, this type of testing is costly and consumes valuable time. The benefit to performing 100% incoming inspection is confidence in the parts placed into inventory. This type of inspection is performed on parts received in very small quantities or when difficulties with a particular part are commonplace.

Incoming inspection equipment for extremely high-volume testing is available widely. This equipment takes many forms, but the intention is to facilitate the rapid and accurate comparison of a shipment of parts to physical and electrical specifications. Where the volumes of each shipment are not as high, lower speed equipment is often employed. In instances where extremely low volumes are received, the incoming inspection equipment can be operated manually, although manual inspection generally is not economical.

Incoming inspection may be implemented as a "process-gate," simply allowing parts that meet specifications to pass into inventory and discarding parts that do not meet the specifications. In this case, little or no information is captured about the parts and what specifications have failed. In situations where shipments contain relatively low part counts, or where parts are purchased from a distributor who may have mixed components from several manufacturers, this type of inspection is sufficient.

Where shipments customarily contain a large quantity of each part from a single manufacturer, more intensive record keeping is in order. Information collected by a customer at incoming inspection can be valuable to the supplier as a means of detecting subtle process errors, particularly those errors relating to quality assurance. Well-managed manufacturing processes seldom allow process errors to escape undetected, but since large-volume purchases involve a more intimate relationship between the supplier and customer, the information fed back to the supplier is as much a matter of courtesy as it is necessary to the continued quality assurance equation. The types of information collected include the number of parts that fail to meet each specification and the actual measurement of the parameter. The information allows the supplier to analyze the severity of, and possibly the causes leading to the errors detected by the customer. The apparent specifications of a part may change after the part is inspected at the factory due to a number of variables, including inadvertent mixing of components, mislabeling of components, and damage during storage or shipment.

Damaged components are common where insulated-gate field-effect transistors are employed. Static discharge is the primary source of damage, due to the extremely high resistance between the gate and channel of these devices. The high resistance and inherent capacitance between the channel and gate allow large charge voltages to accumulate, and eventually punch through or are across the insulating layer between them. As the arc travels between the electrodes in the device, it damages the microscopically thin insulating layer by creating a hairbreadth tendril of conducting material. This conductor is poor, but significantly reduces the resistance of the insulator layer and produces a profound detrimental effect on the performance of the device. The manufacturer of a device takes every precaution to safeguard the part after it passes quality assurance tests, but damage may occur anyway.

Other sources of damage in transit are excess humidity, corrosion, heat, vibration, physical shock, and impact. Many of these factors cause visible damage to the shape, size, color, or texture of a part. Some of the factors produce damage, which is internal to the device or purely electrical in nature, and thus not detectable by visual inspection alone.

In-Process Inspection

Inspection during assembly of a unit is used to detect several classes of errors:

- Components damaged while in storage after incoming inspection
- Errors of assembly or damage from earlier stages in manufacture
- Errors in assembly at the current stage

As explained above, parts may be damaged physically or electrically while in transit or storage. High-volume processes use automatic placement machines to position components on an assembly. Many of the automatic placement machines have the ability to perform simple tests on each component before using it in the assembly process. Manual assembly requires the person performing the assembly to visually and/or electrically inspect a part before using it. This test is not a substitute for the incoming inspection which generally is more comprehensive in scope.

Automatic placement can be affected by the improper loading of parts into the machine, and testing before placement can detect this error. Similarly, a rigid protocol for inspection and testing of each part before manual placement can prevent costly rework at a later stage.

Several types of errors are produced during assembly that are unrelated to the quality or selection of parts:

- Parts omitted during assembly
- Parts positioned incorrectly
- Soldering errors
- Mechanical assembly errors

These errors are most economically detected as close to their source as possible. Fortunately, many of these errors are visually detectable and do not require extensive electrical testing. Soldering errors include solder bridges that electrically short adjacent conductors, and solder voids, which do not make sufficient electrical contact between a component and the rest of the assembly. Mechanical assembly errors are more prevalent on high-energy units than on logic circuits, because of magnetic components that have unusual mounting requirements or intricate electrical connections, and because of the use of heatsinks and fasteners used to mount devices to the heatsinks. An assembly error peculiar to high-energy systems involves the emission of insulators between devices that share a common heatsink element.

Very often an assembly process involving multiple subassemblies is implemented as if each subassembly were a part subject to the same criteria as individual electrical components. Each subassembly is tested and inspected before being placed in inventory, and then given an incoming inspection before being assembled with other components. Inspection of subassemblies reduces the cost of diagnosis and rework that would be incurred if testing was deferred until later in the assembly process. Information about failures is fed back to the earlier stages of the assembly process to allow corrections to be made to the process there, and to solidify a long-term strategy for quality assurance.

An assembly process includes the informational paths found in such a process. Short information paths lead to more efficient correction of process errors. As efficiency of correction increases, costs decline and quality increases. Astute readers will notice that as paths become shorter, more information paths are required to correct a process. An important job of a manufacturing engineer is to manage the flow of information between the various information collection sites and the places where the information is ultimately used. Just as quality is an important facet of a product manufacturing process, the quality of information fed back to previous process steps is important to cultivate and preserve. One should monitor the process as diligently as one monitors the product. Continuous evaluation of the assembly and inspection steps in a process allows another layer of error correction that magnifies the quality control measure within the process, in a sense ensuring the quality of the quality itself. Manufacturing engineers often utilize statistical methods to evaluate their processes. Periodically a few completed units are diverted from the end of the process or removed from completed inventory and subjected to rigorous analysis. The analysis is more extensive than the testing performed in the manufacturing process. Units may be subjected to environmental stress testing (see Chapter 11), destructive analysis or "teardown" analysis in an attempt to uncover defects in the manufacture of the unit.

Ongoing Reliability Testing

Samples of a product may be subjected to accelerated aging or stress to encourage failures. The results of this testing are used to evaluate the process variables that influence the longevity of the product. Among the parameters of a process that can affect reliability are component selection, assembly methods, and thoroughness of inprocess inspection methods.

Pilot Run Feedback

In addition to the local information feedback paths within an assembly process, less formal and less complete information is fed back to earlier steps in the process. The information returned to the designer is one exception to the reduced level of information fed back to earlier process steps. The designer uses the process information to evaluate whether the product design requires modification. When a new process step, new equipment, or a new process sequence is employed, the manufacturing engineer needs to evaluate the process in much the same way that the designer evaluates the product design. Corrections to a process generally are made before the process is employed in high- or medium-volume production for reasons of economy and simplicity. When a high-volume process must be modified, long delays and high costs result.

Every manufacturing process, regardless of its maturity, must be monitored and corrected when opportunities for improvement exist. This so-called fine tuning allows small improvements and changes to be made, however, major changes to processes as a result of monitoring are rare.

Bibliography, Standards of Interest

1. MIL-M-38510: General Specification
2. MIL-STD-883: Test Methods and Procedures for Microelectronics
3. MIL-STD-202
4. MIL-R-22087
5. MIL-R-22684
6. MIL-R-10509F
7. MIL-R-83401 for resistors (among others)
8. MIL-C-11015/20 for capacitors

6

Planning Production Testing

Test Strategies

A strategy is a stated plan of action for meeting defined goals. A power supply test strategy is a stated plan of action for testing power supplies to meet certain defined goals.

The purpose of a power supply test strategy is to create the correct focus around the test issues, for those who are in any way involved with the purchase and/or manufacture of power supplies. In it is defined "the business" and a set of metrics (or measurement criteria) to measure the processes of today along with all of the associated costs, and then define the needed process with the projected costs associated with it for newer products. Finally, the definition of the logical steps (or plan) that need to be taken in order to move from the process of today to the new process in a logical and cost-effective manner. Each step of the process needs to be examined not only to determine its effectiveness, but also its proper place in the process.

It is most important to consider all factors of the business, including: product reliability, cost of goods sold, inventory levels, cost of capital, volumes, product mix, quality, and customer requirements, as they are directly affected by the test process. Each of these factors need to be looked at and understood, in light of how they affect "the business" of today and tomorrow. Plans made need to consider the product mix today and for the future. As volumes, costs, and technologies all change, strategies must keep a step ahead, plotting a course into the future.

Each of the following items must be considered during the process of creating your own test strategy.

- Sales Projections
- Current Production Test Processes
- Long Range Test Process Strategy
- Planning Process Changes:
    * Upgrading current processes
    * New processes ⁻
- Projected Production Costs
- Current Production Processes
- Planning Process Test Tools
- Planning Process Types
- Planning Process Tool Selection
- Planning Military Test Processes

Define the Problem

The problem is usually much larger in scope than it may initially appear. The problem facing every company today is: how do you build products that are competitive, reliable, and still make a profit! A test strategy cannot stand alone by itself, but is a part of the overall business strategy, and must link itself with each of the other facets of the company's or corporation's business plans and strategies.

The job of realistically defining the problem is best performed by a team. A team consisting of persons from at least the following corporate areas:

- Marketing
- Design engineering
- Research and development engineering
- Purchasing
- Inspection
- Manufacturing
- Manufacturing engineering
- Sales
- Service
- Applications
- Facilities
- Corporate management
- Accounting or Finance

Remember that if you don't consider the entire problem, that is, the big picture, then you cannot understand the significance of test issue relative to corporate goals and strategies. Metrics, or measurement criteria, can be used to help identify where you are relative to your goals.

Define Business and Metrics (Measurement Criteria)

Define the Business

Power Products

Identify the power products that are to be manufactured for the next 5 years. Larger corporations may even want to do a 7 year plan.

Identify products that will use or consume power products over the same time span. Scope the size of those products roughly and update the plan as new data becomes available. This data then can become a part of the 5 or 7 year plan for the power products group or division.

Technologies

Determine what technologies will be used in each of the products to be manufactured. Planning technology changes in manufacturing can become a positive event rather than a catastrophe when they are planned carefully well in advance with all team members contributing to the plan.

Volumes

Determine the projected annual volumes per year for the next 5 or 7 years for each product or product family that will be manufactured. Marketing forecasts from the marketing department or from external consultants will be helpful.

Forecasts

Obtain a copy of the marketing forecasts for products to be sold, including new products not yet introduced. Obtain as much data as possible on these new products, even though some may still be the gleam in someone's eye.

MTBF Goals

Determine the MTBF goals for each product or product family to be manufactured.

Time to Market

Estimate the required time to market for each new product introduction.

Define the Metrics

Dollars per Watt

Total price divided by the product of the output volts and output amps. For multiple-output power supplies, sum each output's product of the Volts and the Amps.

Material to Value Added (VA) Ratio

The cost of materials divided by the total value-added dollars, that is, the total direct labor (including assembly, test, etc.) plus the cost of materials.

Cost of Materials as a Percentage of Total Price (Cost)

The cost of materials divided by the total price of the power supply.

Cost of Materials and Manufacturing Overhead as a Percentage of Total Price (Cost)

The cost of the materials and the associated manufacturing overhead, expressed as a percentage of total price, will be an indicator of the percentage of the price, which is expected to be profit (before taxes).

Watts per Cubic Inch

The total watts (VA) of output divided by the volume (product of the length, width, and height) of the power supply.

MTBF

The Mean Time Between Failures (MTBF) can be measured (for older products) or calculated from MIL Standard 217B or some other applicable method, where specified.

Efficiency

Efficiency is the maximum output total VA divided by the RMS input VA. A true RMS reading meter must be used because of the nonsinusoidal waveform of the input current. Off-line switching power supplies are worse than most power supplies due to high third, fifth, and seventh harmonic content in the current waveform.

Ratio of Total Direct Labor to Test Direct Labor or Assembly

Direct labor divided by test direct labor is a ratio. Direct labor is the labor actually involved in producing a product, whereas indirect labor is other labor not involved in the producing a product directly. Assembly, test, solder, and repair are all examples of direct labor. Supervision, management, and engineering are all usually considered to be indirect labor. Either will give an indication of how much direct labor is required for test relative to the total direct labor in a product. Be sure to be consistent in whichever you choose.

Goals

The following list of suggested goals are for your consideration as you start this project. All of the goals are driven by the need to reduce costs as well as improve reliability.

High Process Step Yields

By increasing step yields, work-in-progress (WIP) time is reduces while process step throughput is increased. The reduction in WIP time results in reduced direct labor content of the product while the increased process step throughput reduces the total amount of capital equipment that must be purchased or built, or, the required capital investment for test.

Reduce Direct Labor

Direct labor is a major contributor to product cost. One way of reducing product cost is to reduce the direct labor required to manufacture and test the product. For older products it is sometimes advisable to consider the economics of redesigning a product to make it more manufacturable and more testable. By pushing problems back in the cycle to the point of origin, less labor will be required to find and remove them on a repetitive basis. This requires a close coupling with good communication between manufacturing engineering, incoming inspection, and test.

Reduce Skill Level Required

Reduce the average skill level required for testing. Technicians are expensive and require more technical training to remain proficient at diagnosing and repairing defective products. Operators and repair persons that do not require the same technical know-how are more cost effective in a high volume production environment. It is often easier to hire and train an unskilled worker than it is to hire skilled technicians. However, you cannot just substitute one for the other without modifying the process so that the skill level required is reduced.

Enhanced Product Reliability

By finding the fault that would have occurred in the customer's hands, product reliability is improved. A test process that improves the MTBF of a product allows for less testing at lower costs. Pushing problems as far back in the cycle as possible will reduce costs and increase reliability. A product that is built right the first time is less expensive to build, the cost of repair is avoided, and the product will exhibit a higher MTBF in the customer's hands.

Reduce Operator Training

Reduced operator training results in lower training costs which is usually attributed to overhead, and also results in increased operator availability. An operator that requires less training allows the use of trained operators for that function, rather than expensive technician level operators. This also will reduce somewhat the direct labor content, measured in dollars, of the product. Once again, this requires putting the 'smarts' into the test and manufacturing processes.

Reduce Footprint

Reducing the footprint of test equipment by making it smaller means that it will require less floor space, which will make it possible to produce more product per square foot of plant. This will reduce the cost of capital, namely the buildings, and thereby reducing overhead.

Reduce Idle Work-In-Progress

WIP is inventory on-line. It takes space on the manufacturing floor and requires capital equipment for handling, all which contribute to the cost of ownership. It would be the best possible case if all of the WIP were being worked on at all times. The total time that a product is sitting idle on the manufacturing floor, without any work being performed on it, is a waste of time and money to the corporation. This includes "log-jams" at any process step, even test.

Feedback to Production to Increase Quality

To increase the quality of the products that are shipped to customers, while reducing costs, requires the feedback of useful data from the test process to the manufacturing and engineering processes in a timely fashion. This feedback can help identify particular problems that affect quality and cost of the product. The concept is to push problems further back in the cycle, hopefully to the point of origin, where the problem can be solved rather than tested and removed later. The ideal situation is where all the problems have been solved by the parts vendors and the parts used in production are 100% good. To even start this approach requires a very close coupling between manufacturing, engineering, purchasing, and the vendor base. Working with vendors to achieve the required quality levels is a must.

Better Documentation

In order to improve designs, manufacturing processes, and test processes, the manufacturing engineer needs hardware and possibly software tools that provide him or her with better documentation of the problems, faults, yields, and process step times, etc. so that problems can be quickly traced to the source. Given that data, timely and logical decisions about product changes can be made in a cost-effective manner.

Better Management of Data

Data can only be managed when it is reliable and available. Without proper data, management cannot hope to manage the test process or its yields. All data must be believable and be traceable from a known source at a known time.

Automation

Automation, while not a goal itself, may be considered where high-volume products are being produced. Automation has always made its biggest gains in high-volume of single products; recently, however, automation of high-volume mixed products with some similarity has been making great progress. Production runs with lot sizes as small as one in a mixed production environment can be very profitable.

Business Issues

NOTE
It cannot be stressed strongly enough that you will need to get the assistance of your company financial people and corporate management, to understand these issues and work with you.

Capital (Investment Analysis)

The cost of capital as well as the amount of capital that is available to be spent for equipment are key issues. You will need to show why your plan is more economically sound than some other plan or project being proposed, or why your plan or project is worthy of consideration for funding at the level you have proposed. Two of the important items to consider are the return on investment (ROI) and capital lifetime.

Payback Period

The payback period for equipment is the period of time it takes for the savings that the equipment yields to pay for itself and its installation. Payback periods can run from a few months to several years depending upon the type of investment being made.

Production Capacity

For a tester, one key issue is the production capacity for the process using that machine. This is defined in units per shift or some other relative terms.

Idle WIP

WIP is really "inventory" with its associated costs of ownership. Materials that are idle on a production floor tend to be moved around, bent, broken, misplaced, soiled, contaminated, or just plain get in the way. These materials may consist of partially assembled products, printed circuit boards (PCBs), excess raw parts, etc.

Inventory

Inventory requires floor space, costs money to purchase, and is depleted by spoilage. Excess inventory required to support a manufacturing process that may have bottlenecks, whether they be in test or elsewhere, is wasteful.

Capital Lifetime

The useful lifetime of a capital asset must be estimated or calculated. Items due consideration when considering capital asset lifetime include: wearout, technology changes, obsolescence, capacity, and maintainability.

Direct Labor

Direct Labor is defined as the labor directly applied to assemble, test, and ship a product. Operations such as assemble, inspect, move, test, diagnose, repair, or box, are all examples of direct labor. It does not include supervision, management, or engineering functions.

Indirect Labor

Overhead functions such as supervision, management, design engineering, and production engineering that do not directly contribute to the production of a product are indirect labor.

Cost of Utilities

In some locations, especially those that are off-shore, the cost of electrical power may be very high and may therefore become a significant part of product costs. Because power supplies consume a fairly large amount of power during test and even more during burn-in, the cost of this utility is an important consideration today is some situations. In locations where air conditioning is present on the manufacturing floor, power supply testing and burn-in can add significant levels of heat load to the building. In some cooler climates, this heat could be utilized to lower plant heating costs. It is possible to offset to some extent the high cost of power by utilizing various types of solar power or other energy efficient means in processes. The practicality of each situation must be determined along with good business judgment when making these types of decisions.

Return on Investment (ROI)

Return on investment (ROI) is a key metric in choosing various test tools that require capital expenditures. However this must not be the only qualifier that is used for guiding a selection. ROI is specified as the length of time to the point until an investment pays for itself and starts making a return or profit.

Ratio of Test Direct Labor to Total Direct Labor

The ratio of test direct labor to total direct labor is one metric that is often used to compare products within a company. On occasion, some manufacturers may share this type of data on a limited basis, especially where they are in non-competitive markets.

Management Information Systems

Some method is required to:

- Transfer data, such as 'FAILED TEST', 'R27 BAD', 'SERIAL NO 3713H9823', or other pertinent data.
- Collect the total number of passes and failures.
- Collect data about the types of failures.
- Count the passes through the test/repair loop.
- Assemble the data into a meaningful and useful format.
- Present data periodically to management for review.

Depending upon the size of the business, either a paper system or an electronic Management Information System (MIS) will be required to give the reliable database that will provide needed data for process and product improvements. Failure to do either type of documentation prevents tracing a problem to it's source, and prevents speedy problem solving or process improvement. Computer MIS systems have the advantage of providing data that can be formed into any desired format for a report by sorting and selecting the desired information. Networking test systems and MIS systems creates real-time information flow to drive error detection to its source.

Make versus Buy for Test Systems

One of the key areas where many companies miss opportunities for cost avoidance is in the area of test systems. Many companies do not take the time to examine this issue fully, and therefore operate on the assumption that the cheapest alternative is to design power supply test systems and equipment in-house. Usually this is a costly misconception. There are several key areas in the issue of make-versus-buy that need further examination.

Allocation of Resources

In most companies, engineering talent is scarce enough without adding more tasks to the list of things to do. The real question here is how many new products will not be developed due to the development of in-house test equipment. If the task is delegated to manufacturing engineering, then the question may be, how many reliability improvement or cost reduction changes could have been completed. To do the job right, takes all the time and effort of a full system design project.

Availability

Many test systems are available off the shelf, or with short lead times. The system designed in-house will almost always take longer, and usually will not be as good, only because all of the problems were not thought out in advance! If the equipment required by the application is not readily available, and if a custom modified design cannot be purchased, only then should the possibility of an in-house design be considered.

Risks

The risk of failure with purchased proven test systems is greatly reduced. A successful project that furnishes the tester with only minor delays may still greatly exceed cost projections. With a purchased system, hidden costs are greatly reduced.

Vendor's Financial Stability

Of key importance is the ability of a test equipment vendor to remain financially stable in today's and tomorrow's business environment. This is especially true for smaller companies that could suffer greatly by dealing with a vendor that has lost the race with rapidly changing economic times. Technology is moving as an ever-increasing rate and some vendors cannot keep up with the pace.

Support

Some vendors are more prepared to support their customers than others. Make sure that the level of support required by your company will be available when needed. Some sales have been made in the past based on promises made before the sale. Perhaps the best indication of what services will be available in the future is a listing of what services are actually in place today, and how present customers feel about the response to their requests. Some of the support issues are the following:

Training

Consider the cost of training operators and maintenance personnel, as well as the cost of additional training over time as needed and as people leave and are replaced.

Maintenance

For small companies with only a few testers, all maintenance may be performed under contract from the vendor, if that service is offered.

Upgrading

Upgrading commercial equipment by the original manufacturer is usually much less painful that upgrading equipment designed and built in-house, for many reasons.

Replication

As the business grows, it is much easier to purchase additional testers rather than replicate an in-house design.

Documentation

Tester vendors usually do a much better job of documentation and documentation maintenance, and usually for a much lower cost.

Total Cost

When the total costs involved are all considered, the Buy option usually wins out, unless equipment is not available to meet the specific need. The Buy option can also win out for some military products where security is the prime consideration.

Obsolescence

Whether purchasing or building a piece of test equipment, obsolescence must be planned for at the time of purchase. Estimate the useful life period for the equipment and then periodically review the following questions:

- Is the projected end of life realistic?
- When will the need for replacing the tester with newer equipment occur?
- When will newer equipment present enough cost avoidance to mandate the obsolescence of the older equipment?

Decision

If your decision is to design and build your own test system, then consider carefully the process and the total costs involved. We will now examine the entire concept of choosing a tester to meet your needs, whether such a tester is purchased or built in-house.

Technical Issues

Non-Typical Test Process Steps

Some of the typical process steps may not be commonly thought of as part of the test process, however, they are points where the parts that will become your product are tested, and must be addressed in the interest of increasing product reliability and decreasing costs.

1. **Vendor Test**—While not located in your plant, understanding the testing process steps that are used by vendors will help you focus on the areas where quality assurance testing is needed. This testing may vary from full testing to no testing depending upon the strategies adopted by the various vendors involved.

2. **Third Party Test** (Test Vendor)—While still not located in your plant, understanding the testing process steps that are used by test vendors will help you focus on the areas where quality assurance focus is needed. This testing may vary from full testing to partial testing depending upon the strategies that have been adopted. It is essential that timely feedback of data occur from third-party vendors if it is to be of any real value. The goal is to identify the source of errors and correct the error at the source.

3. **Incoming Inspection**—Incoming inspection is necessary for several reasons: to check the vendor quality level, to prevent bad parts from getting into production, and to use it as a feedback tool to the vendor to help him improve his quality level. In very high volume operations, the concept of vendor process qualification, rather than vendor product qualification, can reduce the level of testing at Incoming Inspection. The goal should be 100% good parts delivered to production a day before they are needed.

In-Circuit Test

In-Circuit testers are usually processor controlled systems that have some capability of detecting process faults. It is important to remember that their only purpose is to detect manufacturing process faults. If bad parts are detected at the in-circuit tester, that is all right, but it is not its purpose. Process faults may be any of the following:

- Missing part
- Wrong part
- Inserted incorrectly, or wrong polarity
- Part broken during handling (or during insertion)
- Bent, broken or misshapen leads
- Opens and shorts
- Solder shorts
- Poor connections

Remember that although the In-Circuit Tester may find some bad parts, its purpose is really to find process faults. There are some circuit configurations where out-of-tolerance parts cannot be measured accurately. For instance, a 1% tolerance part, when in circuit, may be measurable only to 7% in a particular circuit configuration. This may be a sufficient test if it assures you that the part is the correct part. If, however, a similar part is used that has only a 2% difference in value, then additional steps must be taken to confirm the correctness of the value of these parts. Most of the components can, however, be checked to tolerance during in-circuit test, and data about failures should be given to the incoming inspection people so that they can increase testing on spot problem parts. Even if some parts cannot be tested to specifications, the test to a wider specification may be sufficient to identify process faults such as wrong part, missing part, wrong polarity, etc.

Types of Testers

There are many types of in-circuit testers for various applications:

- Analog
- Analog / Digital (mixed signal)
- Digital

Of these testers, only the first two are applicable to power supply applications. For both of these types several technologies are available commercially. They are discussed below.

3-Wire In-Circuit Testers

Three-wire In-circuit testers consist of a source wire, an input wire, and a guard wire. The purpose of the guard wire is to try to null the effects of the other circuit paths caused by the other components on the board. These systems are not usually accurate enough when testing the values of components found in power supplies.

4-Wire In-Circuit Testers

Four-wire In-circuit testers have an added guard sense wire. The total is four wires, consisting of a source wire, a sense wire, a guard wire, and a guard sense wire. All of the wires are scanned under software control, and can be programmed to any desired configuration. By making multiple measurements, this system yields the capabilities of six-wire systems. The four-wires systems can be connected for making Kelvin measurements in very low impedance circuits. For more information, contact GenRad at: http://www.genrad.com.

6-Wire In-Circuit Testers

In six-wire In-circuit testers there is a source wire, an input wire, a guard wire, and a remote sense wire to each of the first three, for a total of six wires. Again, the purpose of the Guard wire is to try to null out the effects of the other circuit paths caused by the other components on the board. The purpose of the three remote sense wires is to null system inaccuracies caused by the resistance of the wires, connections, and switching relays. While this system of six wires appears to have the most accuracy for low impedance devices and circuits such as those found in switching power supplies, some vendors use a six wire bridge connected via a three wire switching matrix. Six-wire systems so connected are as useless as a three-wire system for testing power products. Six-wire systems are offered by several vendors. The biggest difference between the six- and four-wire systems is the software used for system control. If the system software is very user-friendly and provides the required options which make a task feasible, the four-wire system can be more effective in providing the required testability for power systems. Additionally, four-wire systems, such as offered by GenRad, of Concord, Ma, can make more accurate measurements due to lower noise than is present in six-wire systems. This is because they use two of the wires twice, and the noise is canceled out. Four-wire systems are the best choice for power supply boards and like products.

8 and 10-Wire Systems

There are also eight- and ten-wire systems, which are normally used by the National Bureau of Standards (NIST) or other similar organizations. Today the cost of these systems places them far beyond the area of justification for any manufacturing operations.

Making a Choice

In order to determine which tester will meet your needs, several steps need to be taken. First, data is collected about each of the testers you will be initially considering. This data (in the form of specification sheets, vendor data, etc.) will be used to create a comparison matrix to show which of the vendors will have a chance at meeting the needs. Second will be an evaluation of a real power supply board, or a board containing your actual most difficult circuits all together, which each of the finalists will test and provide data on how successfully they tested the product. A less expensive way of qualifying vendors might be only a schematic of a real product, which each vendor uses to generate, via an Automatic Test Generation (ATG) program, a sample program and a listing to what level each part is tested. Be aware that there are several drawbacks to this approach, and use extreme caution if taking this route.

Authors Choice

In general, for power supply applications, the six- and four-wire analog in-circuit testers will give the greatest coverage due to their ability to measure components with very small resistive and reactive values. A four-wire analog in-circuit tester such as the GR2280/2281 may be the best choice for your application, depending upon the component values used in your products. The user interface for this machine is one of the best available. GenRad can be reached at, 300 Baker Avenue, Concord, Massachusetts 01742-2174 or at 1-800-4-GENRAD. Some believe that six-wire systems are slightly more noisy in their approach to measuring a component and that the four wire system cancels some of that noise. Also, testers that are not networkable and cannot share data with management information systems will not be of much use in the near future.

Advantages of In-Circuit Testing

The advantages of using in-circuit test over no testing, visual inspection, or straight functional test and diagnose include:

- Find process (assembly) errors without diagnosis
- Reduce scrap from burned or damaged products
- Increase throughput
- Decrease direct labor costs
- Increase product reliability
- Find bad components
- Reduce Direct Technician Labor costs

Disadvantages of In-Circuit Testing

The disadvantages of in-circuit testing exist especially in low-volume situations, and include:

- High capital cost (volume dependent) and therefore possibly not cost justifiable for low volumes
- Requires some programming expertise

Matrix of In-Circuit Testers

Here we will examine a method for comparing the various vendors and models of In-Circuit testers including an in-house design if that is a consideration. Usually it is not, owing to the complexity of the in-circuit tester. Table 6-1 has been provided for your convenience. The matrix may include such factors as:

- Vendor's name
- Model number
- Machine cost
- Operation costs
- Startup costs
- Field service availability
  * In your plant
  * Mail-in module (board) repair
- Customer training
  * In vendor's plant
  * In your plant
- CPU type
- Networking
- Modularity
- Population
- Total population
- Shipments this year
- Shipments last year
- Instrumentation
- Specifications
  * Measure resistance
  * Measure capacitance
  * Measure inductance
  * Measure analog ICs
  * Measure semiconductors
  * Guarding
  * Specification of test needs
  * Loop testing
- Expandability
- Years in business
- Technology
- Fixed or moving
- Automatic test generation (ATG)

Table 6-1.   Form: Matrix of In-Circuit Testers

MATRIX OF IN-CIRCUIT TESTERS				
MODEL	GenRad	HP	Zehntel	Other
Machine cost				
Operation costs				
Start-up costs				
Field Service:				
In-plant				
Mail-in repair				
Training and location				
CPU type				
Networking				
Modularity				
Instrumentation				
Specifications:				
Measure Resistance				
Measure Capacitance				
Measure Inductance				
Measure Analog ICS				
Measure Semiconductors				
Guarding				
Special Test needs				
Loop testing				
Expandability				
Years in business				
Total Population				
Shipments this year				
Shipments last year				
Shipments in preceding year				
Financial position				
Technology				
Software User Friendliness				
Automatic Test Generation				
Software Debug Tools				
Other vendors considered were: _____ .				

Functional Printed Circuit Board (PCB) Test

Several problems can arise with functional PCB testing, especially where particular PCBs are not completely functionally packaged, that is, where a PCB is part of a power supply system and does not contain an entire function on a single card. Some designs, where a single supply may be spread over several PCB's, are much more difficult and costly to functionally test adequately. Testing these boards usually requires the inclusion of more complex circuitry into the tester or fixture than the usual loads and sources. In performing functional tests, there are several methods that can be employed. They are discussed below.

No Test

Although this is certainly an option, it would require extremely high levels of quality for components and similar levels of confidence for the assembly process, to produce reliable product. The option of no test is, therefore, not a real option to most companies.

Manual Test

The most common type of functional testing of power supplies is done with manual "handbox" testers. These testers are generally low cost, and low-complexity testers are used for both "no go" testing as well as for troubleshooting bad boards. As the number of products increases, the number of manual testers increases until it is difficult to find space for them all. Documentation is more expensive on a number of smaller testers. Making changes to compensate for changes in products can become very expensive, especially if different revisions of a product are to be shipped.

Semiautomatic Testing

Semiautomatic testing was an intermediate step in the evolutionary growth of testing. It was much more prevalent in the sixties and early seventies and was primarily used with early military systems and equipment. This method generally requires significant capital investment, but usually has little payback for most commercial applications. However, a few in-house designers still use this approach. The real drawback to using this test topology is that it must be customized for each product that it tests, and generally cannot be cost effectively designed to met the need of products not yet designed.

Automated Testing

Automated testing of power supply boards and units in high volume applications is usually the most cost-effective method for testing products. Testing usually involves a complete test of specification of inputs and outputs, including signals, clocks, sense, and control lines. Because the board under test may not be an entire product, writing specifications and tests for the board can be very complicated and time consuming. Simple fixtures can be stored for each product manufactured, and used each time a batch of that type needs testing. Being fully automated, a usable database can easily be collected that will yield data that will enable feedback to Assembly, Incoming Specification, Manufacturing Engineering, Design Engineering, and Management. The purpose of the feedback is to give visibility to problem areas, so they can be worked on. Automated testing also will increase throughput, and will decrease the total direct labor contribution to the product, as well as providing other benefits.

Functional Unit Testing

The purpose of functional unit testing is to test the unit against a set of specifications for inputs and outputs. In the power supply business world, inputs are usually AC or DC power, and perhaps a few control signals, and the outputs are usually DC power of different voltages and currents with perhaps a Real-Time-Line-Clock signal and some AC—DC OK signals. There are several ways a power supply can be tested against its specifications, such as manual and automated unit testing.

Manual Unit Testing

Again, manual box testers are used here with great frequency. They are low cost, small, and can be developed rather quickly. They do have some drawbacks which were mentioned earlier, as well as being difficult to calibrate, and difficult to maintain all testers of the same model calibrated the same.

Automated Unit Testing

Automated unit testing of power supplies in high-volume applications is usually the most cost-effective method for testing power products. Very simple fixtures can be stored for each product manufactured, and used each time a batch of units need testing. Being fully automated, a usable database can easily be collected that will yield data that will enable feedback to Assembly, Incoming Specification, Manufacturing Engineering, Design Engineering, and Management. The purpose of the feedback is to give visibility to problem areas, so that they can be worked on. Automated testing also will increase throughput and will decrease the total direct labor contribution to the product, as well as yielding other benefits.

Burn-In and Environmentally Controlled Stress Testing

> Products are burned-in to detect infant mortality failures before they can happen in a customer's piece of equipment. This essentially is a components specification/incoming specification problem. Therefore burn-in will not be discussed here.

The purpose of Environmentally Controlled Stress Testing (ECST) is to detect and correct process faults in power products. Some faults will not show up until the unit under test has reached operating currents, voltages, and temperatures. These faults may be caused by either process problems or by parts problems. The important task is to not only find them before the product is shipped, but to use the data on the cause of the failure to make process changes to prevent the problem from occurring again. Several methods that can be used for environmentally controlled stress testing, and each has its own advantages and disadvantages depending upon the particular application for which the power supplies are to be used. These methods usually include a combination of the following:

- Room Temperature
- Elevated Temperature
- Frigid Temperature
- Temperature Cycling
- Full Load
- Nominal Load
- High Line - Lo Line
- Power Cycling

Although there are many ways that a product could be environmentally controlled stress tested, Table 6-2 attempts to characterize the different requirements and the type of testing that could be used. Your requirements may differ significantly.

Final Unit Test

Final unit testing is usually a repeat of the unit test that was performed before burnin. This test is the very last chance available to detect faulty

Table 6-2. Typical Environmental Stress Conditioning

	Aerospace/Military	Commercial	Consumer
Room Temperature	No	No	Yes
Elevated Temperature	Yes	Yes	Maybe
Frigid Temperature	Yes	No	No
Temperature Cycling	Yes	Yes	No
Nominal Load	Yes	Yes	Yes
No Load	Yes	Yes	No
Shorted Load	Yes	Yes	No
High/Lo Line	Yes	Yes	Maybe
Power Cycling	Yes	Yes	Maybe

units and repair them prior to shipment. Typically, one of the following methods is used to perform that testing:

- Manual Test: Using manual box unit testers to test to specification
- Automated Test: Using an automated tester to fully test to specification

Bad Board Diagnose

Diagnosing faulty boards requires some intelligence either in the form of a technician or in the form of automated diagnostics. Both have their place depending upon the needs of the particular business.

Manual Diagnose via Test Technician

Manual diagnosis is usually done with a manually operated test box and a technician who has been trained and has acquired some experience with the particular unit and is thus able to be a proficient diagnostician.

Automated Diagnostics

Automated diagnostics are performed by a nontechnical trained operator. Diagnostic programs take considerable skill and effort to write. They are usually written by an engineer who can do the necessary calculations and predictions to measure circuit and component performance.

In-Circuit Tester

Testing using the in-circuit tester is performed by a nontechnical trained operator. Most of the programming is done automatically by the tester, but enhancements may have to be added by a technician or manufacturing engineer.

Bad Unit Diagnose

In this case, diagnosing the bad unit to the proper subassembly is usually done via the information gained from the previous unit-test results. If the engineering personnel designed and packaged the unit within functional blocks, the task of diagnosing to the PCB level is usually simple. If however, the packaging design was not done along functional lines, the task of diagnosing to a board level may take considerable skill and training on the particular product. Either way, the goal here is to identify the bad board, replace it, and get the unit tested and shipped.

Bad Board Repair

Repairing the faulty board is usually done by either a repair person trained in repair activities or perhaps the repair technician. In high-volume operations, the technician will mark either a repair tag or will actually mark the actual parts to be changed. A repair person will then repair the boards prior to re-testing. When using an In-Circuit Tester (ICT) to perform the diagnostic task, the board tag may be marked, the bad parts may be designated, or the failure data may be stored for retrieval by an electronic color display system that identifies the faulty components and their location for the repair person. Again, it is the total volume of business that will dictate which of these alternatives is selected.

Typical Manufacturing Test Processes

There are several methods of characterizing manufacturing test processes. The one used here merely counts the number of test steps in the process. For each sequence of steps, various configurations can be utilized, however, due to limited space we have chosen some at random. We suggest that you take the time to develop a number of different combinations that may be more suited to your application than the ones discussed below.

NOTE
In the examples used here, Quality Control Testing and Ongoing Reliability Testing are not considered to be a part of the test process, although they play a very important role in maintaining quality levels in the power products being manufactured. They must not be overlooked as they are important steps in the manufacturing process.

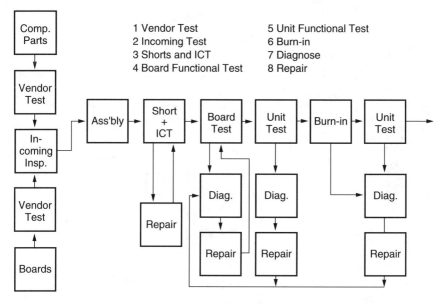

Figure 6-1. The Eight Step Process

The EIGHT Step Test Process

In this power supply manufacturing test process configuration, eight types of process steps are utilized. This type of configuration (see Figure 6-1) is representative of processes in place at several large power supply manufacturing operations in this country today. This type of process is not the most cost-effective approach to manufacturing test operations, but it was not chosen to be. In fact, it was probably not chosen at all, but grew with the particular business until it reached its present state.

The SEVEN Step Test Process

Figure 6-2 shows the next logical step in process development. The seven step manufacturing test process is just the same as the previous example, except that the vendor test step has been eliminated, or rather ignored. This is perhaps typical of most American companies which tend to forget or ignore the vendor's test capabilities. In any of the configurations shown, the vendor tests should not be forgotten, as it can be a real benefit in the ongoing battle for quality products at reasonable costs.

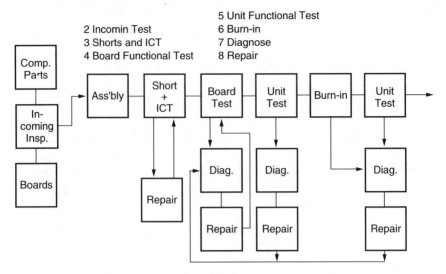

Figure 6-2.   The Seven Step Process

The SIX Step Test Process

The six-step manufacturing test process is just the same as the example above, except that incoming inspection has been removed along with vendor test. For many small companies, incoming inspection seems to be a luxury. They depend upon their ability to detect serious problems during manufacturing and test. Due to their smaller volumes, this is less of a dollar problem. It can impact overhead costs and the cost of goods sold should a component reliability problem arise, as they do from time to time.

The FIVE Step Test Process

This five-step power supply manufacturing test process is perhaps one of the most representative of the type found in many small- and medium-sized companies today. The basic steps of board functional test, unit functional test, burn-in, diagnose, and repair may be arranged slightly differently, but the steps are present.

Another futuristic manufacturing test process that also contains four different steps is shown in Figure 6-3. It gives you an idea of the wide differences in test processes that can be developed using the major basic steps.

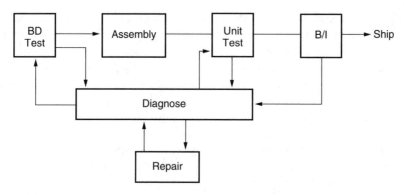

Figure 6-3. The Five Step Process

The FOUR Step Test Process "A"

This configuration as shown is Figure 6-4 makes several assumptions as follows:

- A BOARD ASSEMBLY/ MODULE is defined as a printed circuit board assembly that has been built, wave soldered, and that has already

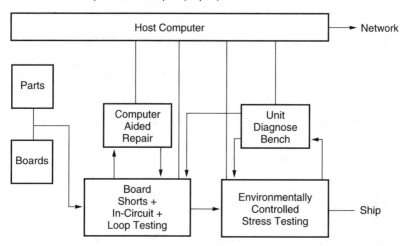

Figure 6-4. The Four Step Test Process "A"

undergone all hand operations performed on the assembly. It is ready for the first level of testing, which consists of shorts, opens, in-circuit test of components, and loop testing.

- A UNIT is defined as any assembly that may be shipped to a customer. It is a unit that is functionally complete; it may only consist of a PCB assembly or may consist of many PCBs in a chassis. Because it is a functionally complete unit, it can be tested on an input/output tester; a bed-of-nails type fixture is not usually needed for functional testing. The definition of a unit includes spare boards, option modules, or other single boards that may be shipped without integration into a multi-assembly unit.

- LOOP TESTING is defined as the testing of a functional unit as follows: all bias points are biased by external power supplies, the 300 V bus is powered by approximately 40 V, and little or no load is placed on the outputs. This allows the regulator circuitry to regulate the outputs as designed. Loop testing does not apply full stress to the product, but does check that all the loops "play" together operationally.

The following assumption is made about new product designs:

- Products designed by engineering will be about 85% worst case designs. To utilize 100% worst-case designs may not be cost-effective. The goal is to solve the serious problems in design and then fix the others later as they occur.

The following assumption is made about environmentally controlled stress testing (ECST):

- It cannot be entirely eliminated in the near future for power supply products. Process tools **MUST** help reduce the need for environment tally controlled stress testing to the most economically optimum level. To do this, process tools must provide information on failures during environmentally controlled stress testing, as well as provide a guide to the appropriate level of stress testing for each product in near real time.

The FOUR Step Test Process "B"

The four-step test process "B" shown in Figure 6-5 is another very popular test process used by companies that build mostly single board power supply products. The process steps are: unit functional test, burnin diagnose, and repair. This process is used mostly by smaller companies using manually operated equipment. It represents a low capital cost approach but is also

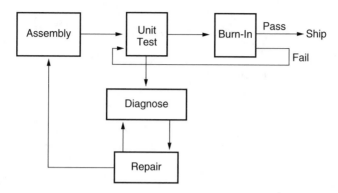

Figure 6-5. The Four Step Test Process "B"

much more labor intensive. For many small operations that ship less than $3 million worth of power supplies this test process seems to be a common practice. The burn-in used in this process is usually short and is accomplished at room temperature, or in some cases at elevated temperature, usually using resistive loads and sometimes power cycling. When the burn-in is in the self-heating mode, some temperature cycling will result from power cycling.

The THREE Step Test Process

Although the three-step test process shown in Figure 6-6 is used by some firms, this could allow a considerable number of defective products to ship. The process faults that can be detected only when a power supply is subjected to full load for a period of time to bring it to full operating power and temperature. This particular process will also require much more repair capability due to the fact that the process faults will not be discovered until a power supply is fully assembled. It must be then partially or fully disassembled to do the repair. Also, when faults are not found before first

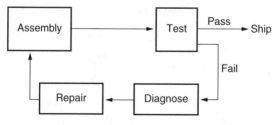

Figure 6-6. The Three Step Test Process

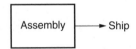

Figure 6-7. The 'No Test' Test Process

power-up, the fault may cause damage to other components and perhaps even damage the PCB of the unit under test. This process is not recommended for serious consideration.

The 'No Test' Test Process

It is theoretically possible that even a "no test" test process could be the correct choice, depending upon the needs of the business. That is, it is perhaps possible to achieve a process with few enough faults that a large number of failures are prevented in the field at first power up. Although this scenario is theoretically possible with complete automation, a quick final automated test is still needed to provide a sanity check on the process. Figure 6-7 shows the "no test" test process.

Manufacturing Test Process Considerations

Design for Testability

For power supplies, testability is really two-pronged issue. The first is the packaging of the topology into standalone functional blocks, and the second is the achievement of worst-case designs so that whenever the correct parts are correctly installed the boards will always work.

Testability issues must be worked with the designer of the power supplies directly. Without testability, the power supply manufacturing test process becomes larger, more complex, and more costly. The issue of worst case designs is more easily addressed for linear power supplies than it is for switching power supplies.

For switching power supplies, businesses who can justify the cost of Computer Aided Design (CAD) tools for worst case design, are in a position where they can obtain higher manufacturing yields than their competitors. This puts them into the businesses where large volumes make them cost effective. In situations where there will not be a good Return On Investment with full CAD tools, using simpler tools and generous engineering practices can help solve producability problems. Even without any CAD tools, the issues still remain the same: packaging of the topology into standalone functional blocks, and worst-case design. They must not be ignored!

Test Time

Test time can be a limiting factor in high volume applications where a production tester is in use full time. The total number of capital dollars to be spent is equal to the product of the test and handling time per unit and the number of units to be tested that month. This time is then divided into the number of machine hours available (with operators) for the month to yield the number of testers needed. One way to increase machine availability is to add shifts per day until a full 24 hour span is covered. One factor that limits full tester capacity utilization is the human element. Operators take breaks, eat lunch, get sick, need training, get tired, take vacations, have bad days, and have all sorts of other problems that will limit their effectiveness. The effective availability of test operators is estimated at 75–85%. Where operator time, not machine time, is the limiting factor, an extra floating operator to fill in for breaks, lunches, and sickness may be required to increase machine utilization.

Test time for most in-circuit testers testing power supply boards is usually in the 60s range, and test time for most unit testers is usually in the 30–90 s range.

Test Coverage

In-Circuit Testers

The issue of in-circuit test coverage for power supply boards (modules) revolves around the number of difficult to measure components on the board. For linear power supplies test coverage can exceed 75%. For switching power supplies however, the coverage can be much less depending upon the complexity of the board to be tested and the capabilities of the in-circuit tester being used. Here the coverage may range from 50 to 80%, depending upon the complexity of the board being tested. With the addition of LOOP testing, coverage can increase to above 99%.

There are a number of companies that manufacture analog in-circuit testers that will meet the needs of many linear power supplies. For switching power supplies however, the needs are much more demanding. To achieve the same level of test coverage on a switching power supply board as on a linear power supply board will require a much better tester, which will be discussed below.

Power Supply Unit Testers

Generally power supply unit test coverage is a function of he capability of the tester and the requirements of the test specification. Coverage of 98–100% is normal.

There are about a dozen power supply unit testers available from vendors, with wide-ranging capabilities. Many of them are real competent and can meet the test needs of a power supply manufacturing organization. The key factor here is making the match between your needs, and what each vendor offers.

Test Process Step Yield

This is really the most important issue. Test process step yields must be sufficiently high enough, that the next test process step will not choke on the level of failures. If, in a high volume situation, 99% yields are required at each next test process step, and it takes 25 process steps to attain that goal, then the required number of steps is 25! With automation, and very high volumes, the goodness of the process step is the most critical issue.

Political Issues

Identify the Problem

One of the biggest problems is that everyone sees each problem from different angles. If the problem can be identified and everyone agrees that the correct problem is being addressed, then the first step toward mutual agreement on a solution has been accomplished. Every organization has two power structures; the "formal" and the "informal". Be aware of both of them and success will come much easier.

Identify the Players

The most important task is to identify correctly all the involved persons that can be supportive of the project and its implementation. The list of players may include persons from the following areas:

- Corporate management
- Finance
- Marketing
- Sales
- Design engineering
- New product engineering
- Manufacturing engineering
- Test engineering
- Purchasing
- Inspection/quality assurance
- Component engineering
- Manufacturing
- Production control
- Test
- Service/field service, etc.

"Not Invented Here" Syndrome

Some individuals believe that unless they've invented it, it isn't good enough, isn't advanced enough, doesn't address their problems, or a myriad of other such reasons. Therefore, many companies end up in the business of building power supply testers when they could have easily purchased excellent systems from established vendors. One positive response to this type of thinking is to present the both the positive and negative benefits of buying off-the-shelf versus built in-house. When the corporate view of allocation of resources, time to market, capital expenditures, total project costs, hardware support, software support, and documentation, are all understood, this type of thinking tends to disappear.

Propose an Acceptable Solution

Careful consideration of alternative solutions will help to find the solution that will be most acceptable to all parties concerned. If the solution presented is considered to be radical or unrealistic, the chance of success will be diminished. One way to arrive at an acceptable solution is to inform the interested parties, over a period of time, of the results of the various stages of the project, even when it is in the very early formative stages. Many people will accept new ideas when in the very early stages they were asked for ideas and suggestions, and have been asked to review the plans along the way. If some of their suggestions are incorporated, then the whole plan is partly theirs! Of course they will support it!

'Buy-In' by All Parties'

Buy-in by all parties means that all (or nearly all) of the concerned parties have agreed with your strategy, its goals, and its implementation. Depending upon the management style of the organization, this could be extremely easy, very difficult, or anything in between. This is the point where you must be a good salesperson, contacting each interested party in advance to get advice in this type of a project. One of the worst things that can happen is to start an implementation plan and suddenly find out that there is no mutual agreement on goals, strategy, or an implementation plan.

Replacing Emotions with Facts

For some it is very difficult to separate emotions from facts. When such a case presents itself, it becomes necessary to step back and get a clear understanding of the facts. Avoid using words such as: my, our, your, I, or we and instead concentrate on using words and phrases like: the project, Corporate needs and goals, a concept, and, one method. By getting cooperation from all the players involved in the power business at an early stage, in the project, emotional conflicts can be avoided.

Sample Strategy Document

A sample corporate power supply test strategy document is located on the following pages. Modify it, change it, add to it, and fill it in to make it fit your needs and the needs of your company.

Power Supply Test Program Goals

Here the goals of the program are communicated to the reader. For example, here are two goals that typify the goals of a program such as this.

1. To ensure that the company's long range analog and power supply test needs are defined, and a plan put into place to fill those needs.

TITLE PAGE

POWER SUPPLY TEST STRATEGY

for

YOUR COMPANY NAME HERE Corporation

Today's Date

by

Author's name here

COMPANY NAME HERE

CITY AND STATE HERE

Revision and Date

Table 6-3. Product Volume Forecast

PRODUCT VOLUME FORECAST				
Products	Last Year	This Year	Next Year	Next Year +1	Next Year +2
ALPHA	1200	1000	800	200	0
BETA	4700	5000	4400	3700	1700
DELTA	0	500	3300	7900	7700
PI	0	0	500	900	1500
EPSILON	0	0	0	10	25
OMEGA	0	0	0	0	5000

2. To evaluate, select, and introduce new systems and processes into the company for the testing of analog and power supply: boards, modules, and units that meet the corporate goals and strategy.

Products: Today and Tomorrow

The products of today and the future must be shown, with forecasts for volumes by quarters of a year, to show the scope of the business at hand and the direction of growth. This will allow you to forecast the testing volumes and the processes and equipment to support them.

Table 6-3 is only an example of what could be done. More extensive data should be presented, with quarterly or monthly figures, in both dollars and units, to give the reader a more complete picture of the total business. Additionally, data can also be presented on some of the other metrics such as Dollars per Watt, since these figures will help show to what extent you have studied the problem and done your research.

Definitions and Assumptions

To be certain that your strategy is clearly understood, list your assumptions and definitions of key words and concepts that will be used in the strategy. This will go a long way towards elimination of confusion or misunderstandings. Here are some examples of Assumptions and Definitions:

Definitions

Board Assembly/Module Assembly

A PCB assembly that has been built, wave-soldered, and on which all hand assembly operations have been performed. It is ready for the first level of test, which could be any or all of the following: shorts, opens, in-circuit test of components, and loop testing.

Unit

A unit is defined as any assembly that may be shipped to stock or be dock- or customer-mergable. It is a unit that is functionally complete and may only consist of a PCB assembly itself. Because it is a functionally complete unit, it can be tested on an input/output tester, and a bed-of-nails type fixture is not needed for functional testing.

Loop Testing

Loop testing is the testing of a functional unit as follows: all bias points are biased by external power supplies, the 300 volt bus is powered by approximately 40 V, and little or no load is placed on the outputs. This procedure will allow the circuitry to regulate the outputs as designed. Loop testing does not apply full stress to the product, but does check that all the loops "play" together operationally.

NOTE

The '40' V applied to the DC bus may vary from 30 to 90V depending upon the application.

Design Assumptions

- Products designed by engineering will be about 88% worst-case designs. To do 100% worst case designs may not be cost effective. The goal is to get the serious problems in design, and then fix the others later, should they occur.

NOTE

The 88% figure is a randomly selected number and not intended to suggest a value for your operations.

- Burn-in will be eliminated for power supply products. The use of burn-in for elimination of infant failures in power supply products is not cost effective in the commercial marketplace, these problems must be worked out with the vendors.

- Process tools MUST help reduce the need for Environmentally Controlled Stress Testing to the economically optimum level. To do this, process tools must provide information on failures during all testing, including Process Maturity Testing, and Ongoing Reliability Testing, as well as providing a guide to the appropriate level of Environmentally Controlled Stress Testing for each product in near real-time.

Today's Test Process

Our goal here is to show a diagram of today's test process as it exists for current products. See the example below.

Today's Power Supply Manufacturing Test Process

Database for Today's Test Process

The next step is to create the database by gathering all the applicable data.

NOTE

This data may perhaps belong in an appendix rather than here, but it would be good to refer to it.

Screen Time

The actual time to test or screen a board or unit, not including the time to load and unload a fixture, make connections, or any other similar operations not directly associated with the testing of the product.

Diagnose Time

The actual time to diagnose a board or unit, not including the time to load and unload a fixture, make connections, locate equipment or any other similar operations not directly associated with the testing of the product.

Yield Percent

The percentage of product that pass test as compared to the total tested, expressed as a percentage.

Handling Time

The actual time to load and unload a fixture, make connections, or any other similar operations, except the testing or diagnosis of the product.

Units

Units means numbers of units.

Failures per Million Units

From the data of number of units to be shipped per year per type, and the projected yields, the total number of failures per million, or a lesser number if desired, will give real impact to the size of the problem.

Graphs of Today's Test Process

Today's Manufacturing Test Process Inventory

Table 6-4 is intended to show product inventory time for each of the process steps. An alternative table could be one showing product inventories (in dollars) for each of the process steps.

Direct Labor

Table 6-5 is intended to show the direct labor contribution for each of the process steps. An alternative graph or chart could be one showing direct labor costs for each of the process steps.

The major difference between these two is in the area of burn-in where the direct labor time show is only for the handling of the unit to put it into and take it out of the burn-in chamber.

Migration of Test Process

The block diagram in Figure 6-8 shows how today's process can be made to migrate or evolve into the process of the future. Notice that much of the detail of both of the processes is left out, showing only the overall concept.

Table 6-4.   Today's Power Supply Manufacturing Test Process Inventory

| TODAY'S POWER SUPPLY MANUFACTURING TEST PROCESS INVENTORY | | | | |
| Time per 100 Boards | | | | |
SHORTS & ICT	BOARD TEST	UNIT TEST	BURN-IN	UNIT TEST
100 w 243	199	0	1804	202
500 w 275	447	280	4404	310
1000 w 275	555	620	9803	564
Minutes	Minutes	Minutes	Hours	Minutes

Table 6-5. Today's Power Supply Manufacturing Test Process Direct Labor

	SHORTS & ICT	BOARD TEST	UNIT TEST	BURN-IN	UNIT TEST
TODAY'S POWER SUPPLY MANUFACTURING TEST DIRECT LABOR					
Times per 100 boards					
100 w	243	199	0	270	202
500 w	275	447	280	400	310
1000 w	275	555	620	700	564
	Minutes	Minutes	Minutes	Hours	Minutes

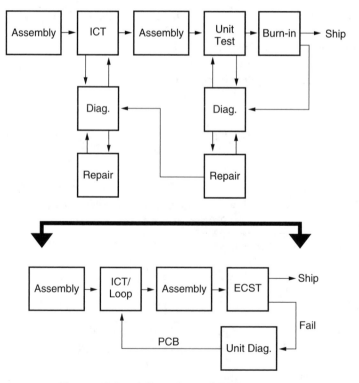

Figure 6-8. Migration of Test Process

Evolution of the Power Supply Test Process

Proposed Test Process

The proposed test process is now shown in more detail, with all the process steps clearly identified. Clearly identify any unusual or new process steps, so that the reader can plainly understand the process from this document alone. It is important that the first view of the new process does not confuse the reader, but shows the overall concepts without excessive detail. This first peek should be kept simple. More detail can be added once the basic concepts are grasped. See Figure 6-9.

Database for the New Test Process

The database below was created as an example of what a typical database might look like. The numbers contained in this example were picked from a hat. This database is another candidate for an appendix, rather than in the main body of the strategy.

Graphs of the New Test Process

The purpose of the graphs is to depict the results of the new test process visually. An not so old saying goes, "A picture is worth ten thousand bytes." At this point in the strategy, the reader hopefully has a mental picture of what the test problem is today and is relating that to the future. At this

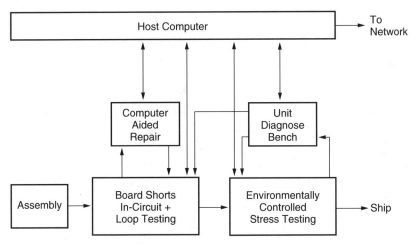

Figure 6-9. Proposed Test Process

Table 6-6. Test Database for New Test Process

The Power Supply Manufacturing Test Process Database

(per 100 Units)

	Shorts, ICT and LOOP test			Computer aided repair			Environmentally Controlled Stress Testing			Unit Diagnose		
Unit Size	100W	500W	1KW	100W	500W	1KW	100W	500W	1KW	100W	500W	1KW
Screen Time	2.0	2.4	2.7	3	5	7	4	6	9	1.5	2.0	3.8
Diagnostic Accuracy	99	97	93	—	—	—	—	—	—	99	98	97
%Yield	90%	90%	90%	99%	99%	99%	95%	95%	95%	99%	99%	99%

point, the reader starts to get a first glimpse of the solution. The secret is to keep it simple, especially on the first pass. See Figure 6-10.

The New Power Supply Manufacturing Test Process

Table 6-7. New Process Product Inventory

New Process Product Inventory (per 100 Units)			
Shorts, ICT and LOOP test	Computer aided repair	Environmentally Controlled Stress Testing	Unit Diagnose
100 W 341	12	213	3.5
500 W 363	21	703	5.0
1000 W 373	21	2204	7.8

Table 6-8. New Process Product Direct Labor

New Process Product Direct Labor (per 100 Units)			
Shorts, ICT and LOOP Test	Computer Aided Repair	Environmentally Controlled Stress Testing	Unit Diagnose
100 W 354	12	100	3.5
500 W 363	21	230	5.0
1000 W 373	21	500	7.8

Today's Test Process versus the New Test Process

A comparison of today's process with the new process is made to show the projected savings that can be made. The data from this comparison may be included in the introduction to attract the reader's attention. Only one table or graph is shown here, but data should also be included for the other power supply business metrics (measurement criteria) as well. The projections of

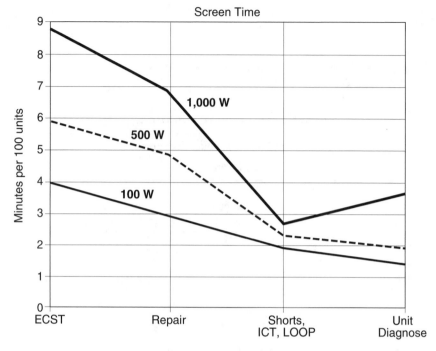

Figure 6-10. New Test Process

direct labor for test as well as the other metrics will help management formulate better plans for the future.

Proposed New Equipment for the New Test Process

Potential Equipment List

- Incoming Specification Tester
- In-Circuit Tester
- Computer Aided Repair Bench
- Environmentally Controlled Power Supply Stress Tester

Attributes of the New Power Supply Test Process

- Higher Product Quality
- Pre-conditioned components
- Shorter time to market

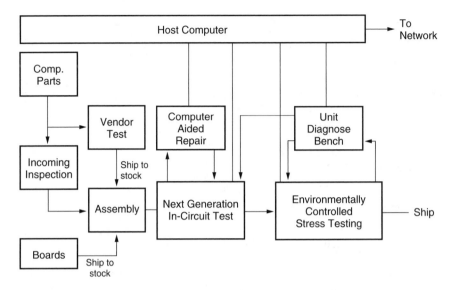

Figure 6-11. A Modern Test Process

Table 6-9. Projected Savings of Inventory and Direct Labor (per 100 units)

Projected Savings of Inventory and Direct Labor (per 100 Units)				
	Inventory		Direct Labor	
Today				
100W	815 Hours		29 Hours	
500 W	2426 Hours		45 Hours	
1000 W	9646 Hours		60 Hours	
Proposed				
100W	210 Hours		15 Hours	
500 W	818 Hours		29 Hours	
1000 W	2367 Hours		42 Hours	
Difference				
100W	605 Hours	74% Red.	14 Hours	48% Red.
500 W	1608 Hours	66% Red.	16 Hours	36% Red.
1000 W	7009 Hours	73% Red.	18 Hours	30% Red.

- Eliminate burn-in
- Lower product costs
- Higher yields

These attributes are a result of the following factors:

- Computer Controlled Process
- Fewer Process Test Steps
- Specified Test Interfaces
- Environmentally Controlled Power Supply Stress Testing
- High Levels of Testability
- Highly Automated Material Handling
- High Quality Tests
- Highly Efficient Tests

Matrices of Available Test Equipment Vendors

In Table 6-10 is an example of the type of paper study that can be performed. Following a paper study, the next step is to try a sample product on the top two or three systems. This can be done in several way:

- Try via demos or a loaner or lease the best system as shown by the paper study, and if it works, buy it.
- Try the best two or three systems with representative product for a final decision on the best machine for the process.
- Evaluate the best two or three systems with a specially fabricated PCB that contains all of the various technologies used in the products to be tested.
- Try the best two or three systems by representative product test evaluation, and evaluation with an specially fabricated PCB that contains all of the various technologies used, for a final decision on which equipment (hardware and software) best fits the needs of the process.

The choice of one of the above methods, or some other qualification method, will depend upon the size of your business and its particular needs. There is no one right way that will fit all companies. The differences in products, size, and philosophy demand that each firm makes the choice that best meets it's needs.

These matrices do not contain lots of detail about each tester, but one could be created to allow for a more exacting comparison of each of the

Table 6-10. Form: Matrix of Power Supply Testers

MATRIX OF POWER SUPPLY TESTERS					
	AutoTest		Schaffner	In-house	NH Research
Cost average					
Fixture + development $					
Start-up costs					
Field Service					
• In each plant area					
• Mail-in repair available					
Training					
• In-factory					
• On-site					
CPU Type					
Modularity 0-9					
Population					
• This years shipments					
• Last years shipments					
AC Source 0-9					
Measurement System					
DC Loads 0-9					
Ripple & Noise 0-9					
Expandability 0-9					
Technology 0-9					
Auto Test Generation 0-9					
System Burnt-in (Hours)					
Years in business					
Dun & Bradstreet Rating					
Overall Rating					
Other Vendors Investigated _____					

test system. Depending upon your needs, each of the vendors offers equipment and/or capabilities to assist you in the power supply testing task. Use the Matrix in Table 6-10 and 6-11 to help evaluate each system, then make a choice. The vendors listed are the major vendors. There are others that may be:

- New companies, or companies with new offerings.
- Unknown to the author, as the author does not know about all existing firms.
- Companies that offer test systems that do not come close to meeting the requirements for testing the power supply products at hand. Such firms will not be listed.

Proposed Projects

At this point, a list of proposed projects, proposals for which are located in the appendices, is included. These projects might be similar to the examples given:

- Power supply in-circuit tester
- Environmentally controlled power supply stress tester
- ICT diagnostic programs

A project summary may help to show the timing expected for the projects that you are proposing, and any progress that has been made to date.

Test Technology Projects

Present the projects here in Table 6-12 showing major deliverables, clearly identifying the month and year of completion. This example will give you an idea of the format.

In the project plan you must communicate to all who read it, the reason for the project. Describe the need for the project fully. Then go on and describe or even outline the steps to evaluate the various testers and how the decision will be made in choosing the tester. Show how the process will be developed in cooperation with all the players that need or want to be involved. In the final report, describe how you met the goals and plans and why you missed some dates, if you did. Be very careful not to place blame on people, but on the processes that were used. You may also want to

Table 6-11. Form: Matrix of In-circuit Testers

MATRIX OF IN-CIRCUIT TESTERS					
Model		GR	HP	ZEHNTEL	OTHER
Machine Cost					
Operation Costs					
Start-up Costs					
Field Service:					
	In-plant				
	Mail-in repair				
Training & Location					
CPU Type					
Networking					
Modularity					
Instrumentation					
	Measure Resistance				
S	Measure Capacitance				
P	Measure Inductance				
E	Measure Analog ICs				
C	Measure Semiconductor				
S	Guarding				
	Special Test Needs				
	LOOP Testing Capability				
Expandability					
Years in Business					
Total Population					
	Shipments this year				
	Shipments last year				
	Shipments preceding year				
Financial position					
Technology					
Software User Friendliness					
Automatic Test Generation					
Software Debug Tools					

Table 6-12. Test Technology Projects

Test Technology Projects			
Major Project	Forecast Date	Scheduled Date	Actual Date
Environmentally Controlled Power Supply Stress Tester			
Projected Plan			
Evaluate Concept			
Procure Tester			
Develop Process			
Final Report			
Next Generation Power Supply In-Circuit Tester			
Project Plan			
Evaluate & Purchase			
Develop Process			
Final Report			

suggest in this final report how a next project may be done differently and better, with reasons why.

Funding Requirements for Proposed Projects

At this point it is necessary to communicate the cost of implementing this strategy. This section must be written with the help of the financial department, from whom input into the strategy has been solicited from its beginning. Funding requirements should include:

- Major Deliverables. List here the major pieces of equipment and major milestones that are to be achieved.
- Project Cost Total. Show how money will be budgeted for the project.
  * Project Expense Cost Total
  * Project Capital Cost Total
- Project Human Resources requirements. Show the needed personnel to support the project, and when each will be needed. Included projections for current employees, new-hires, consultants, and contract help.

- Project plant requirements. Show the needed floor space, and any changes needed to the plant, such as: movement of production lines, more electricity, more air conditioning, and additional needs like vacuum lines and pumps.
- The Alternatives. Show the negative effect of not doing the project and its effect on production and profitability.
- Related Projects. Here you will want to show any ongoing or proposed projects within the company, that may aid in the successful completion of the strategy and the projects needed to implement it.
  - * Fixturing Development
  - * Computer-Aided Repair Bench

Implementation

- Review One-on-one With Each Player. Each of the players that was identified earlier should be contacted, and met with on a one-on-one basis, to review the entire strategy and project plans. Address minor differences as quickly as possible, "on-the-spot" is preferred.
- Present to Corporate Management. Make a presentation to the appropriate corporate management. Because of the size of the data involved, send them copies a week prior to the meeting so that they can become familiar with the material. I recommend that you approach each of the persons who will be attending the meeting well in advance, and review your presentation with them, answering any questions. When all parties have a clear understanding of your strategy, they can be supportive of your case.
- Identify a sponsor (get funding). In a large corporation, identify a sponsor who will fund your projects and back your strategy through implementation. In other corporations and smaller companies, this may be the appropriate manager with the budget to support the strategy.
- Start Implementation Phase. Once approval for the strategy and projects is received, start right in with them. Make it a team effort, involving all the players that you have been working with all along.

Conclusions

First, achieve understanding of the whole problem. Get the big picture clearly in focus by gathering data to obtain a clear understanding of all of the parts of the picture.

Next, gather the goals of the corporation, of your group, and of power supply test, to help define the task at hand. Be sure to work with all the applicable parties in the corporation.

Be sure to consider all of the pertinent business issues and work with all those who have the necessary expertise to correctly define the issues relative to power supply test.

Work with engineering, vendors, and others to clearly understand all the technical issues relative to power supply testing of today and the future. Identify each of the test process steps for the current process as well as any planned process, and understand how these steps can be utilized testing power supplies.

Examine the current manufacturing test process, and all the other possible manufacturing test process configurations. The choice of the best test process is one of the most important tasks at hand. Be bold in your approach.

Examine each of the possible manufacturing test process steps for ways that it can be made better, or combined with other steps to reduce the total number of steps.

Understand the political issues and prepare to work with people from each of the groups of disciplines.

Create the outline of the strategy and start to fill it in. You may find it helpful to use the suggested format, with some modification to suit your needs.

Implement your strategy, with the aid of your team of experts, and advisors. Help each team member to participate in the area of his or her expertise.

Suggested contents for such a document are given here for you to adapt, change, and modify to suit your needs.

1. Title page
2. Assumptions and definitions
3. Today's test process
4. Database for today's test process
5. Graphs of today's test processes
6. Migration of test process
7. Proposed test process
8. Database for new test process
9. Graphs of new test process
10. Comparison graph: today's test process vs. the new test process
11. Proposed new equipment for the new test process
12. Matrices of available test vendors and equipment
13. Proposed projects
14. Funding requirements for proposed projects

15. Related projects
16. Implementation
17. Review one-on-one with each player
18. Present to corporate management
19. Identify a sponsor (get funding)
20. Start implementation phase
21. Conclusions.

Bibliography

1. Crandall, Earl, "Creating a Power Supply Test Strategy," a one-half day seminar presented at Powercon10, March 23, 1983, San Diego, CA.

7

Low-Volume Processes

Types of Processes

Low-volume manufacturing processes are limited in the viable options for both assembly and testing. The driving factor for the application of a particular type of testing is most often cost-related. The options open include manual test processes, semiautomated test processes, and automated test processes. Because of the limited number of units produced in most low-volume processes, setup costs for many options simply cannot be justified as they constitute an added expense to each unit produced. In a few situations where quality or performance are paramount, the issue of cost may be diminished or ignored, and the number of options open in selecting process steps increases. A process is most often considered low volume by virtue of the number of units processes. A low-volume process is one where the rate at which units are produced is low, even though the process may be used for years to produce thousands of units. Manual assembly and testing processes are used by the majority of low-volume manufacturers because of cost constraints. The cost of full manual test processes quickly rises with any appreciable volume.

Manual processes are more time-consuming than those using automation and labor costs are a prime contributor to the expense of manual process steps. As the amount of automation used by a process step is increased, the amount of human labor per step is decreased and the costs per unit decrease while the overhead costs in terms of process development and setup charges increase. Figure 7-1 illustrates the relationships between the amount of automation and both the cost per step (recurring) and the setup cost (nonrecurring). The cost per step includes equipment and labor costs and any expendable material or utility costs. In most cases, the setup costs are entirely nonrecurring and may be amortized over the process volume. Higher volumes allow the nonrecurring cost per unit to be reduced. Figure 7-2 is a graph showing the effect of increasing the number of units produced upon the total cost of manufacture. This graph is simplified, and is not intended to show the additional beneficial effects of volume purchasing discounts or the economy of scale in inventory and asset utilization. Even without these effects, the cost per unit decreases somewhat as volume increases.

Choosing the Correct Process

Manual Testing

The lowest-volume processes are totally manual. The instruments and tools used in the testing stations are operated by people. Units are connected to the test instruments by people, and the test instruments are controlled and

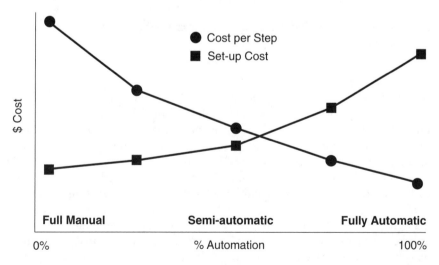

Figure 7-1.   Relationships between the Amount of Automation, and Recurring and Nonrecurring Costs.

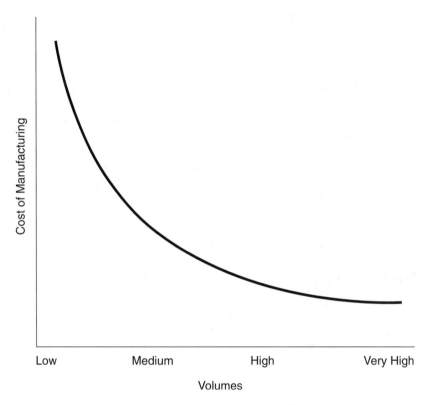

Figure 7-2. Effect of increasing the number of units produced versus total manufacturing cost.

observed by people. The people used for these types of test stations fall into two categories, skilled and unskilled. Skilled test station operators understand the operation of the units being tested and the operation of the test equipment, while unskilled operators usually follow a set of step-by-step instructions to perform the tests without knowledge of the details of the unit or of the test equipment. Small companies commonly employ skilled operators to operate the test equipment for several reasons, again economic in nature. A small company may not have sufficient test operations to require a single full-time test operator, so the test operator is generally skilled in another facet of the manufacturing process. In addition, small companies usually produce limited numbers of each type of unit, which complicates the training of unskilled test operators and makes creation of step-by-step test instructions more difficult. Skilled operators have several advantages over their unskilled counterparts: a skilled operator more easily compensates for unexpected behavior in the test apparatus or the unit under test. Skilled operators perform tests using complicated instruments more rapidly than unskilled operators. Skilled operators can detect abnormalities in the test procedures that would ordinarily escape detection by someone unfamiliar with the internal operation of the unit or test equipment. Few unskilled operators are able to diagnose the failure of a unit without detailed diagnostic procedures. Manual testing is not without problems. As mentioned above, the labor costs for manual testing will exceed the labor costs for testing with some degree of automation. Considerable variation in quality can result where the human element is involved, and that usually mandates additional quality control processes to measure variations in the product and reduce the differences to tolerable levels. Additional quality control steps increase the cost of manufacturing a product, often canceling any economic benefits gained by choosing manual testing over more automated techniques. The quality of a product is directly traceable to the invariability of the product parameters. Reducing variability improves quality.

Here quality is defined as "the conformance to expectations of the user," therefore, high conformance (i.e., less variance) is high quality.

Variability introduced by humans can come in a variety of forms. The accurate adjustment of controls and interpretation of instrument readings differs from one person to the next. Factors of fatigue, disinterest, distraction, and confusion also influence the performance of human elements in a process. "Never buy a car manufactured on a Monday or a Friday" is a warning offered in jest to many automobile purchasers that voices a concern about the quality of human labor at work on the auto production lines just before or after a weekend. While the quality of an automobile manufactured one day may or may not differ from the quality produced any other day, the common warning shows an awareness of human variability by average consumers. A wise process engineer understands the repeatability of human labor, and designs a process that can accommodate and correct the variability that humans introduce. Manual tests require very little planning before they are implemented. The test engineer must select appropriate instruments and interconnecting equipment. Where unskilled labor is used test instructions must be prepared. These preparations do not consume very much time in relation to the planning needed for test processes with larger percentages of automation.

Semiautomated Testing

Significant reductions in process variability are possible with the judicious application of automation to a manufacturing process step or test process step. Automation may be mixed with traditional manual operations to achieve the goals of reduced cost and less variability when compared to purely manual process steps. Automation may be targeted at those parts of the process where variability affects the product most, leaving the rest for the human element. In automation-assisted processes, the human being performs those operations that are either less important to the quality of the product or those tasks that are inappropriate for automation due to the complexity or delicacy required. A human being may be required to connect a unit to the test setup and then command the test equipment to perform each step in the sequence. The automatic equipment configures itself for each step and presents the result of each measurement to the operator for evaluation. The person decides when and if to continue to the next test in a sequence of tests based upon the results of previous tests. The person is assigned to a decision-making role, rather than the tedium of following a step-by-step set of instructions for performing a test.

Another common practice is to have a computer direct a human being in the testing of a product. The computer program makes decisions based upon readings fed to it by the human being, and the computer commands the individual to perform the steps that would normally be read from a step-by-step instruction guide. The human being acts as the hands and eyes for the computer. The computer program decides whether the parameters resulting from a test are acceptable or not, and chooses the next appropriate test step. There remains considerable opportunity for operator induced error in semiautomated processes. A person may follow instructions imprecisely, or may inaccurately report a measurement to the system. The human element is prone to many of the same sources of error as those affecting manual processes. By limiting the interaction of the human within the semi-automated process, variations in the process may be minimized.

Every test has four potential outcomes. A unit may pass a test or it may fail. Two other possibilities exist: a failure may be detected where none exists, or a failure may escape detection. The first two consequences are expected and are the basis upon which all testing is predicated. The case where nonexistent errors are detected may not adversely affect product quality, but does increase costs and reduces production quantities while the units are analyzed and re-tested. In the final scenario, where failures escape detection, process quality suffers. Human intervention in a test reduces the test effectiveness by increasing the probability of an error being introduced. Human intervention also makes the test process more vulnerable to improper assessment of the test results. One viable goal of semiautomated testing process is to attempt to minimize all effects of human action, and make special efforts to avoid letting errors escape detection. Redundant testing is one method of reducing operator error. Multiple test stations may be used in sequence to reduce the effect of any single operator, or multiple tests at one station may be used to correlate the results of one test with another. Either technique achieves the desired effects. Semiautomated testing requires more planning than does a manual test operation. Some custom equipment may be required, and test instructions are developed in a more formal manner so these preparations require additional care and time. The use of automated equipment is less forgiving than purely manual processes, therefore much of the additional time is consumed by activities anticipating the problems with a process that will be less flexible than a process operated by a human.

Automated Testing

The most effective way to control test effectiveness is to remove humans from the day-to-day process altogether. Automated testing is subject to the same four outcomes mentioned before, but the process tolerances can be maintained much more consistently than processes with some manual intervention. Repeatability is the strong suit of automated processes. This repeatability is the primary weakness of automated testing, because an automated process has no mechanism to respond to unfamiliar stimuli. An automated process demands consistency in both the product design and the spectrum of problems encountered. Those situations that deviate from the realm of the expectations of the test designer are opportunities for reduced test effectiveness. An automated process requires considerable planning and foresight. The test designer must ensure that the test processes are complete and that they do not unduly stress the units being tested. Because automated tests are performed in a more rapid sequence than humanly possible, stored energy from a previous test may be discharged in a subsequent test through an unexpected path rather than dissipating harmlessly, The components in the discharge path may consequently be damaged or destroyed. The test designer must work closely with the circuit designer to avoid these circumstances by careful selection and review of potential test methods and sequences. Modern in-circuit test generators and simulators allow the prediction of test completeness as a part of the simulation. Armed with this test coverage information, the test designer is able to manually intervene in the test generation procedure to improve the completeness toward the goal of 100% coverage. (Note: **The goal *is* 100% test coverages, right?**) Functional test generation for nonlogic circuits is considerably less mature than its logic counterpart. In logic circuit testing, very limited conditions may exist at each node in the circuit, and each node has clearly defined source and sense paths. It is more complex to analyze analog circuits for test generators since the acceptable values for each node are not limited to a handful of exclusive states, but may have an infinite number of possible values, each of which may affect adjacent nodes differently. This complexity decreases the likelihood that complete test coverage will be achieved, and increases the intricacy and cost of the test generation software. For low volume processes, test generation may be more manual than automatic because automatic analog test generation tools are generally not economical, and many of the techniques for improving test coverage require considerable experience with the test generation software. The cost of automated test is shifted away from the individual test to test preparation, and in many low-volume processes this may cause automated testing to be more expensive than manual testing. Although fully manual processes are less expensive to develop, they vary more than automated or semiautomated processes. This variance requires additional testing and inspection to restore confidence in the product quality. Fully automated processes are the least variable, but also require considerable investment in the process development and analysis phases, not to mention the higher initial cost of automated over manual equipment. The decision of how much automation to use and where to use it are not simple choices. The following factors must be considered for each process step:

- How much variability does a human introduce at a process step compared to equivalent automation of the same step?

- What are the consequences of increased variability of a process step? In other words, does the human produce tolerable quality variance, and (if not) what are the penalties in terms of customer satisfaction or lost sales, or rework costs?

- How much would the process to correct the variability cost?

- How much would automation to replace the human element cost? Remember to include equipment costs, maintenance costs, expendables costs, utility costs, and the additional process development costs.

- How will the preparation required by automation affect the time to create a viable manufacturing process, and the time to market for the product?

- Would partial automation be beneficial at this time or is total automation the only alternative to manual testing?

- What are the costs versus benefits of partial automation?
- Is equipment to automate a process step available widely, or must it be custom developed (which may add time delays and increased cost)?
- What are the labor cost savings of introducing automation at a process step by reducing the need for skilled test operators?
- Does automation or partial automation make sense based on expected number of units being produced?

8

Medium-High
Volume Processes

There are primarily two sizes of manufacturers: low-volume and high-volume suppliers. Generally, companies and product lines expand from small start-ups, so a very successful low-volume plant must make a number of transitions on the way to becoming a high-volume operation. The strategies that are used successfully by a low-volume manufacturer are less effective when the number of units manufactured per month increases more than a certain limit. Strategies used by manufacturers with very large volumes appear inappropriately capital intensive for a manufacturer with a good low-volume operation. Realistically, most manufacturers are somewhere in between the two extremes of the scale, and always facing decisions on changes as they expand (or contract) overtime. Low-volume processes are characterized by a high percentage of manual operations and are not automated to any large extent. Conversely, the typical high-volume process is automated wherever possible and may require the participation of only one or two people in monitoring and correction tasks on an entire production line. The transition from a labor-intensive to a capital-intensive operation is complicated and expensive, and the largest hurdle for a manufacturer expanding a process. The final goal is not to be a medium volume plant but to be a high volume manufacturer. It is important to keep this goal always in sight during the transition process. With this goal in mind, the manufacturer in transition attempts to minimize the time and expense of expanding, while always looking for ways to maximize profitability.

The relationship of economic viability to volume may be pictured as a cup shaped curve (Figure 8-1). Low-volume plants are viable because they

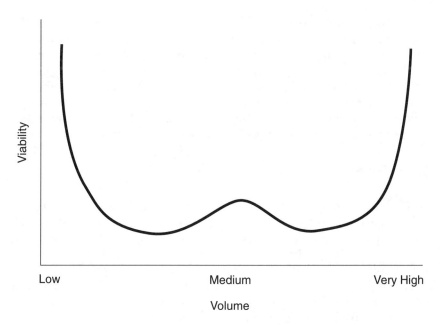

Volume

Figure 8-1. Economic Viability

have low process overhead and low capital costs. High-volume manufacturers are viable because they are able to amortize their considerable investment
in equipment and planning over a large number of units. Once a supplier
moves away from these stable points and begins to adapt elements of the
other style of manufacturing, the costs begin to erode profitability. The
medium volume manufacturer is in the worst spot on the economic curve,
and moving toward either profitable style of manufacturing is a continuing
struggle. The real trick is to match expansion and growth with the increased
capabilities of the high volume-processes and still make a profit. While it
would seem obvious that manufacturers wish to either minimize the time
spent transitioning from a low-volume to a high-volume operation, or to
minimize the depth of the curve, this is the secondary concern where the
long term strategy is concerned. It could be that the plan must include a
period of low profitability or even some loss during the transition period.
The strategy is the long term plan which allows a smooth transition and
greatly increased volumes and, therefore, profits. It is just as dangerous to
underestimate as it is to overestimate transition and its effects.

A low-volume manufacturer may have a difficult task migrating from
the realm of manual processes to the high-volume paradigm of automated
processes. Much of the effort involves learning the manufacturing and pro-

cess strategies and developing the process measurement tools to control a high-volume process. Without careful planning and the correct process control strategies, a high-volume process can go out of control at great cost.

Changes selected to implement a change in process style should be chosen with an eye to their eventual use in high-volume manufacturing. Investment can be justified, in contrast to high-volume processes which are able to utilize automation in all but a few instances, and low-volume processes which generally cannot support the overhead associated with automation.

Lot Sizes

A common technique used to ease the complexities of adopting a new style of manufacturing is to build units in groups of a predetermined number. This simplifies many of the issues of inventory and process control. It is possible to emulate a high-volume process for a short period of time and gain benefits in labor distribution, training, purchasing and equipment utilization.

The appropriate number of units in a lot or block of manufacture is usually determined by the amount of inventory that may be accumulated and the relative cost of the units. Large lots allow the process to spread the overhead costs over more units, but will increase the warehousing costs and tie up capital in inventory until the units are sold. If possible, a manufacturing run of one or more lots may be scheduled after an order is received, thus minimizing the float stock at the risk of being able to fulfill the order within the contract time. Relatively small blocks of manufacture avoid inventory and warehousing difficulties but do little to impact the nonrecurring engineering, capital, and administrative expenses that characterize processes that include any degree of automation. Many highly automated processes allow lot sizes as small as one unit.

While fixed lot sizes allow process engineers to characterize the processes using statistical tools, production requirements often require variable lot sizes.

Employing high volume techniques in an intermittent fashion may allow a broadly skilled manufacturing team to better utilize the human resources. With fixed lot sizes, production managers are able to very accurately assign assembly, test, and inspection personnel when they are most needed at various points in time during manufacture of a lot.

Types of Processes

The medium-volume manufacturing test process varies from one implementation to another, and thus there is no single model that may be diagrammed and discussed. The typical process is a hybrid of both low-volume and high-volume manufacturing styles.

Testing in a medium-volume environment may include automated equipment that is functional such as in-circuit testers, module functional testers, unit testers or units that may attempt to combine these functions. These systems can provide a direct path to a very high volume environment, if the transition strategy has correctly forecast business directions to allow the proper planning of the required types of systems found in high volume plants. The disadvantages include high purchase prices and significant non-recurring engineering expenses, whereas the advantages include very-high-volume capacity per system, reduced operator training requirements, and the fact that these systems will not be obsolete when higher volume processes are contemplated. Generally, the automated systems have the ability to handle a broad range of product sizes and outputs while requiring far less manual labor, and less trained labor as well.

Semiautomated systems assist trained operators in the test, inspection, diagnosing and eventual repair of modules and units. The equipment used for semiautomated processes is less expensive to purchase, and has significantly less ability to process high product volumes. The advantages include low cost of the system and higher throughput than fully manual testing. The disadvantages are that these systems become increasingly inappropriate as volumes increase, that the operators must have a higher level of skill than people automated equipment, there are still considerable non-recurring engineering expenses, and every product slated for these systems must be carefully analyzed to determine if the systems will be adequate for the task.

Manual testing is not a viable option in medium volume scenarios because the capacity of manual testers is too low or the level of quality is too inconsistent to allow statistical process management. Additionally, manual testing requires highly skilled operators and technicians. Automated systems chosen correctly for the task can reduce the level of skilled personnel to half in many cases, volumes being equal. Statistical process management becomes increasingly important as product volumes in crease. Statistics allow the identification of general problems in the manufacture of large numbers of units. A trend toward poorer output regulation might identify a faulty batch of components, a dysfunctional assembly operation, or an inspection step that is not stringent enough. Trends are difficult to discover in low-volume manufacturing because of the percentage of operations that are manual in nature and thus not well controlled or documented, and conversely the constant presence of human eyes, minds and hands is missing in the medium volume plant, so the continuous inspection and correction of process faults must be replaced by something.

Choosing the Correct Process

Automated Testing

Automated testing is very seductive to persons whose sole experience is in low-volume manual processes. Process engineers that are familiar with low-volume processes accept a large number of inspection steps as a method of correcting the highly variable human-controlled process steps. Automated testing offers the ability to perform the inspection faster and more repeatably than is the case when manual testing is used. The setup and programming of automated testers is complex and time-consuming, and therefore expensive. The education expense associated with the learning curve on new equipment is higher and is spread over a longer period of time on automated equipment as opposed to its manual counterparts. The use of automated equipment will inevitably raise costs in the short run but will save money in the very long run. If a plant produces products with very short production lifetimes, the economic benefits of automated testing may never be realized.

Consistency of the tested product is simple to achieve with automated equipment. This predictability may be exploited by enhancing the product's test specifications and using the automated test system to ensure that they are being met. Tight control of specifications is difficult in a manual testing environment, because of the variability between humans in their analytic and monitoring skills. Consistency of testing and statistical process monitoring are two tools that allow the analysis of products to discover opportunities for further test specification enhancement.

Repair and maintenance costs for these systems must be considered or you may have a costly surprise later. Lack of maintenance can lead to poor-quality testing and reduce the effectiveness of the process. Planning for replacement of older instruments and upgrading of equipment is the best way to avoid any surprises in the future.

Semiautomated Testing

Many situations do not produce enough revenue to justify the relatively high overhead and non-recurring expense (NRE) costs of fully automated testing. In these cases, the logical substitute for fully automated testing is semiautomated testing. This method uses human labor and intellect to augment the automated process. The human factor is applied where physical or intellectual complexity of one or more tasks would have most increased the cost of test automation. A design that incorporates adjustable components such as potentiometers will obviate fully automated testing in all but a handful of cases, as automated adjustment systems are extremely expensive and difficult to implement. Semiautomated testing is most often found at the device functional test and unit functional test process steps. Less than full automation of in-circuit testing is prohibitively time-consuming.

The typical semiautomated testing station incorporates a computer or computer terminal, several instruments controlled by the computer, and space for the unit being tested to be connected to the tester. An example is shown in Figure 8-2. More complex stations may include sophisticated cooling arrangements for the device being tested, additional instruments not

Figure 8-2. Semi-Automated test station.

under the control of the tester, and a collection of tools for assembly and disassembly of the unit.

The computer directs the sequence of operations to be performed at the station, and controls the instruments that will stimulate the unit and measure the response. The human operator makes the physical connections between the unit and the tester, operates the stimulus and measurement instruments that are not under computer control, and performs any physical tasks that the tester may unable to perform automatically including adjustment of components.

The operator may be prompted for each action by the computer in explicit language, or may follow a test sequence in a written document and receive prompts that only indicate the applicable step number in the document. In either case, the operator is not highly skilled, and merely provides the manual dexterity or intellectual flexibility inherent in the human.

Semiautomated testers are very similar to fully automated systems, with the primary differences being in the complexity of software used, and the impact upon design alternatives that full automation implies. Semiautomated testers use software that is much less tolerant of variations in the product, but is less complex and is simpler to develop. A manufacturing process that is loosely controlled characteristically contains semiautomated testing, because the manufactured units will vary enough in physical or electrical parameters to make full automation difficult.

Tight control of the design and manufacturing assembly processes reduces the cost penalty of full automation compared to semiautomation. Careful planning of a product and its manufacturing process will allow use of fully automated processes and provide a clear migration path to high volume manufacturing.

Semiautomated Diagnosis

Diagnosis of modules and units is considered an art by many. There are many similarities between electronic diagnostic analysis and the diagnostic analysis performed by medical personnel. The actual error may not be directly observable in many cases, and therefore must be located by examination of symptomatic evidence and deductive reasoning. Deductive reasoning implies a core of experience that will influence decisions about probable sources of the fault and will direct the effort to gather additional clues. Skilled technicians are expensive, especially in a high volume or even medium-volume plant.

Semiautomated diagnosis employs the human operator as a skilled, reasoning entity. In very simple electronic designs, the level of skill and ability to reason required of the human operator is minimal. The diagnosis of complex designs will not be effective without highly skilled operators having considerable experience. Computer programs written to simulate this skill and experience are lengthy and often error-prone. Diagnosis is one step in the manufacturing process that is better performed by a skilled human than by a computer, because of the mental complexity and reliance upon experience in the task.

In semiautomated diagnosis a computer may be used to pare the symptom list or the list of suspected components to manageable size. The human being then examines the evidence, or gathers additional information with manually operated instruments to aid the reasoning process. The human operator directs this process, and uses the computer as a data reduction tool. Note that in semiautomated diagnosis the human is in command, but during semiautomated testing the computer does the directing.

Other semiautomated diagnosis strategies use preconfigured diagnosis stations that have instruments connected to sequencing equipment that is less a computer and more a station with selectable preset conditions. The operator manually selects the appropriate conditions and sets switches on the tester to create the appropriate stimulus and load. This technique requires more physical hardware and manual switching than the computer-based approach, but the costs are comparable.

Semiautomated diagnosis is, at the best, an expensive, labor-intensive stopgap measure. The real goal is to remove the need for expensive direct labor from the process altogether.

Removing the Expensive Diagnosis Process Step

In typical medium- to high-volume test processes there are several steps of interest to us as seen in Figure 8-3. All of the failed modules are sent to a

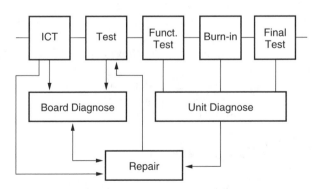

Figure 8-3. Typical Medium-High Volume Test Process

diagnosis area where highly paid skilled diagnostic technicians try to determine the causes of the actual fault. In a typical medium sized plant there could easily be as many as 15–20 highly skilled diagnostic technicians. The cost of this direct labor is charged directly to the cost of the goods being sold, thereby greatly increasing its cost. But this added cost doesn't appreciably add to the value of the product. Some years ago in attacking this very problem, albeit from a different angle, we were able to reduce the skilled technician count from 12 to 6 by merely adding loop testing to the major product in the line.

LoopTesting was developed to reduce the direct labor content in a product, and increase the quality of that product. The goal was and is to make a better product for less actual cost.

Description of LoopTesting

Loop testing is a method of testing power supplies without full power or full loads to the extent that fresh lot yields at the next step approach 100%. For an off-line switching power supply this may mean that the input voltage is raised to 20 or 40 or 60 V, depending upon the application, so that the loops close, and that the output at no-load produces the full output voltage. This can be done on an in-circuit tester (ICT) such as the GenRad 2280/81 series. This means that the ICT replaces the functional tester for modules. Tested modules are then tested at full power at final assembly with very high fresh lot yields. Any module failing loop testing can be re-routed through ICT for diagnostic testing. This lowers test costs.

Implementation of LoopTesting Process

Loop testing can be implemented on an analog ICT such as the GenRad 2280/81 series. First the in-circuit test is run, and if a module passes the ICT test, the loop test is automatically run next. Those cards that fail are automatically run through the ICT test again to point out the failing component or components. Any modules that fail in final assembly are also routed to in-circuit testing for diagnosis. At the start of a new product, each failed module is run through the ICT system until the system can accurately diagnose the fault. Test engineering will want to increase the diagnostic accuracy as quickly as possible.

Cost Reduction Considerations

One way to reduce the cost of power supplies is to remove some of the cost of diagnostics by letting the machine do the diagnosis instead of trained

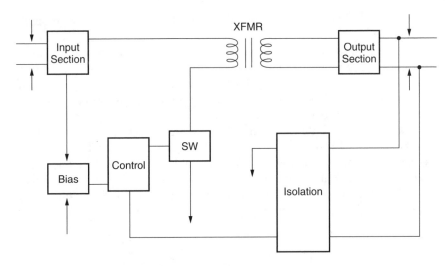

Figure 8-4. Loop Testing Concept

electronic technicians. ICT operators are less expensive that diagnostic technicians, and require far less training.

Quality Improvement

Quality improvement comes as a result of the ever increasing of ability of the diagnostic in-circuit testing and loop testing. This task may become best performed by the product engineer on the manufacturing floor. His or her goal is to drive yields up and failures down. The ideal process is one where faults are discovered as they are made, and corrected at that point. Remember, the goal of the test process is to drive fault detection back to the point of origin in the manufacturing process where they are eliminated. Therefore, better quality products come off the end of the production line.

Partial Automation

Another technique for producing medium-volume processes is the use of partial automation. Some steps are fully automated while others are manual. Those process steps that allow cost effective automation are implemented with full automation, and those process steps for which the planning and expense cannot be justified remain manual. This strategy spreads the transition cost for migration from low volume to high volume over the entire transition period, in contrast to semiautomation which may lead to multiple incarnations of the same equipment as the human element is gradually phased out.

Partial automation retains much of the variability of the completely manual process, but reaps many of the cost of transition benefits of full automation. Each manual process step must be evaluated as volumes increase to determine which steps must be automated and when the transition from manual to an automated step must occur.

The medium-volume operation is a manufacturing plant in transition. The goal is low non-recurring expense (NRE) costs and complete automation with its attendant low incremental cost per unit of manufacture. The strategies of medium volume manufacturing are a means to that end, rather than an end in themselves. Plan for high-volume operation and implement full automation where the engineering costs can justify it.

9

Very-High-Volume Processes

The production of large numbers of devices allows many economies of scale. Computerized equipment operates very quickly, often working faster than the combined abilities of several humans. Completely automated process steps are more consistent than processes in which humans play a part. These conditions lower the cost of manufacturing each unit but the capital expense, planning, and programming of these systems is significant. The nonrecurring costs must be recouped by dividing them over the entire) volume of product produced.

Equation

Unit cost = parts + labor + incremental overhead + (NRE/volume)

Figure 9-1. Equation for UNIT COST.

The parts cost of a particular design are relatively constant, although some minimal advantage may be realized through volume purchasing. The incremental overhead includes packaging, warehousing, warranty costs, and other costs not related to the design or direct manufacture of the device. The labor expense and share of non-recurring expense (NRE) are the two items most affected by complete automation. Highly automated processes are characterized by relatively low labor costs and relatively high NRE in comparison to manual or semiautomatic processes. By generating sufficient volume, the contribution of nonrecurring expense to the product cost can be minimized. The most important consideration is the reduction of labor so that higher volumes produce higher profit, thus firmly establishing the process as a profitable high volume operation. Manual processes are an anathema to high volume manufacturing.

Another characteristic of high volume processes is widely variable lot sizes. The advantages of consistent lot size that were important to medium volume processes are insignificant to high volume operations. Lot sizes may be included several thousand units or may contain only one unit. Just-in-time manufacturing is made simpler by high-volume manufacturing techniques.

A typical high-volume process is shown in Figure 9-2. Modules are assembled and delivered to in-circuit testing (ICT). The ICT step locates open and short circuits, and does loop testing. Faults are repaired and returned for another cycle through the in-circuit tester. The assemblies that leave this step have been tested so that significantly less than 1% of the potential errors in the modules remain. Modules that have completed testing are assembled into units and delivered to environmentally controlled stress testing (ECST). More than 99% of the units that reach this step will operate correctly at turn-on.

ECST lasts several hours or days, and units that fail are removed from

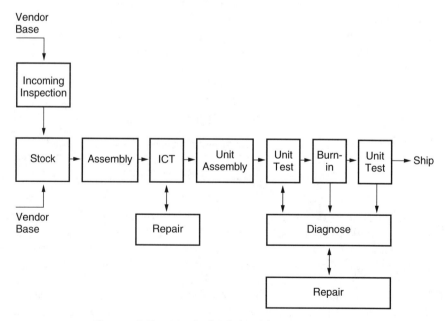

Figure 9-2. Typical High-Volume Process.

ECST for diagnosis and repair. The diagnosis determines which modules in each failed system are at fault. The unit is disassembled and the suspect modules are delivered to the ICT step where they are tested and ultimately repaired. The operational modules from a failed unit may be tested or simply returned to the unit assembly process for inclusion in another unit.

Units that complete ECST are packaged and shipped to the customer. The repair step of this process is the only part of the process that requires significant contribution by human beings, otherwise the other steps may be completely automated. Where an assembly process is under strict control and is well understood, the failures may be so few that faulty modules are simply discarded, eliminating virtually all human labor. A "lights out" automated factory operates without direct contributions of human beings and (as its name implies) does not require factory lighting or air conditioning for the operation of the process. Most completely automated factories employ a few human beings to monitor the process and to intervene when the process goes out of control. A process that remains in control is a prime candidate for lights out automation, as the need for human intervention is dramatically reduced.

Figure 9-3 shows the ideal process in a well-understood manufacturing line. The repair operation is eliminated as faulty module assembly is rare enough to permit discarding failed modules. Likewise, the diagnosis and disassembly steps are missing as the cost of these operations exceeds the benefit of correcting units that fall out of the process. The ECST step is the quality gate for the entire process. Any units that complete the ECST process are considered acceptable and are eligible for shipment to a customer. A process that remains in control is a prime candidate for lights out automation, as the need for human intervention is dramatically reduced.

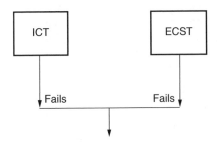

To Manufacturing Engineering or Test
Engineering for Evaluation and then Test
Process Modification/Change.

Figure 9-3. Manufacturing Line Ideal Process.

Choosing the Correct Process

A fully automated process step is considered an "island of automation," and a fully automated production process could be pictured as an 'archipelago of automation.' Each step operates as an independent process, accepting input from other steps, and control from quality assurance sources; it produces output that is as much as product of the step as the finished unit is a product of the factory. In Figure 9-4, each workcell is shown as an independent entity with no ability to communicate with other process steps.

Most automated equipment is able to connect into a system of information interchange, often termed a 'network'. In the network, each tester and process machine reports status and product information. Sophisticated equipment has the ability to evaluate the information on the network and adapt the workflow based upon that information. Networked assembly and test process equipment is the first step in achieving this capability, as shown in Figure 9-5.

A classical networked factory is depicted in Figure 9-5. The master control system provides information about the entire process and allows a single point of control. Data concerning quality assurance issues is available from this system and changes in the process are entered here. The controller at each workcell communicates with the master control system to get test limit information and to report discrepancies. More than one workcell of each type (assembly, ICT, ECST) may be used to increase, volume capability.

Figure 9-6 shows a similar topology. The primary difference here is that the each workcell is controlled independently but has the ability to "snoop" at the process quality information to make independent adjustments. This topology is virtually immune to failures of the central computer, but has no single point of control as found in the classical network. This method allows versatile implementation of each process step without the need to conform to a specific communication or inter face protocol. The distribution of quality information may be triggered by humans, or scheduled on a periodic basis, or it may be available for asynchronous requests from each workcell. Data about the process may not exist in a central repository, but

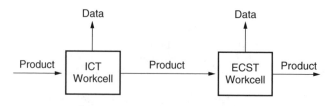

Figure 9-4. An Advanced Process Flow

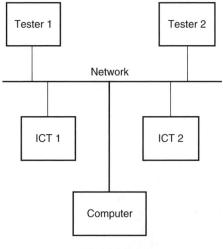

Figure 9-5

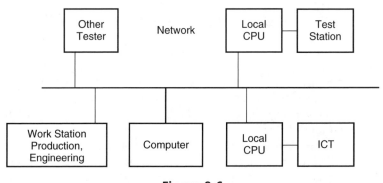

Figure 9-6

may be spread among the various workcells. The task of the Central computer may be to compile reports rather than to control the process.

Computerized Diagnostics

The thought processes that are required to emulate reasoning and deduction are not well understood, so computerized diagnostics are based upon simple pattern matching, or upon rules. A computer has a "dictionary" of faults, and simply looks for a match between the responses gathered from a faulty unit and stored "symptom signatures." An exact comparison triggers the output of a message corresponding to the signature. The message is usually intended for human interpretation, but may be applied to a automated repair station without the benefit of human intellect.

ICTs diagnose by testing every component on a module and evaluating the integrity of circuit traces between components. This method of dividing the circuit into a relatively large number of subcircuits and testing each one thoroughly allows simple fault isolation.

Fault isolation is the primary mechanism of diagnosis. The goal of fault isolation is to reduce the search for faults to a small portion of the circuit. Testers may use several different methods for eliminating a sub-circuit from suspicion, and the sub-circuits that remain after this selection process are considered likely faults. The ideal situation would place each component in a separate sub-circuit so fault isolation would result in a list containing only the component or components that were actually faulty. In-circuit testing approaches this ideal very closely.

Power supplies by their nature are stabilized by feedback, regardless of whether they are linear or switching designs. Feedback 'loops' found in virtually all designs allow cumulative tolerance error on components in the loop to produce circuit failures although all of the components are within tolerance. Many design tools permit worst case analysis and of tolerance, but designs with this particular problem exist. The only way to troubleshoot problems of this type is to perform a loop test, which applies power to the components in the feedback path and tests the operation of the entire loop. Some testers operate the loop in 'closed form' as it is normally used, while other testers break the loop and test the components as an amplifier. Figure 9-7 shows a classical feedback loop, and Figure 9-8 depicts the circuit in various open loop forms.

Testing is more difficult when integrated circuits are used in the design, particularly if access to critical nodes of the circuit is denied by the package. The use of large scale integrated circuits in power supplies has simplified the design of these units and improved the reliability by reducing the number of parts used, but made testing more difficult. In-circuit testers may be unable to evaluate the health of a complex integrated circuit, and this will affect the accuracy and reliability of computer-aided diagnosis.

Manufacturers of some large-scale ICs provide the ability for these chips to test themselves. This 'built-in test' simplifies the task of the test designer, and lowers manufacturing costs. The test system merely triggers the built-in test on a chip and monitors the result. There is no assurance that the built-in test is completely accurate or covers the entire function of the circuit, and there is no reliable way to automatically test the interconnecting bond wires within an IC, so some margin for error remains. It is this margin for error that contributes to the errors that will escape detection at the in-circuit tester.

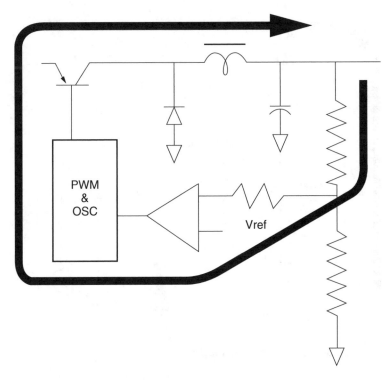

Figure 9-7.   Simple Loop

Another scheme for testing digital ICs at reduced cost is covered by an IEEE Standard by the Joint Test Architecture Group (JTAG). This technique reserves a portion of each IC as a 'test interface', and gives the test designer a way to look inside a complex integrated circuit. This method allows the same interface to push signals into internal nodes inside the IC to evaluate their effect. This 'window' into the nooks and crannies inside of an IC, and reduces the number and duration of tests that must be performed to verify the operation of an IC. These are all digital test techniques and most generally not suitable for power supplies with the very limited amounts of digital logic circuitry.

It is possible to create Analog JTAG test circuitry, and several companies are examining such concepts, but it will still be quite some time before this technique becomes of interest to the power supply designer. Planning is the most important specification of a high-volume process. The goal of low incremental cost per unit may be achieved when careful attention is given to issues of design-for-testability and quality assurance.

In-circuit testers may be unable to evaluate the health of a complex inte-

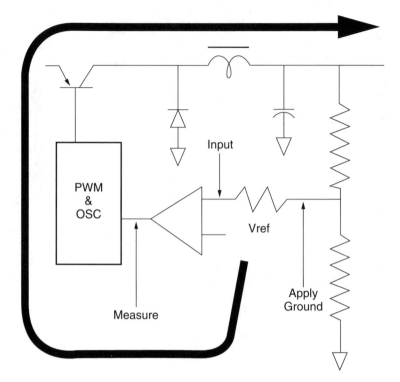

Figure 9-8. Open Loop

grated circuit, and this will affect the accuracy and reliability of computer-aided diagnosis. Manufacturers of some large scale integrated circuits provide the ability for these chips to test themselves. This 'built in test' simplifies the task of the test designer, and lowers manufacturing costs. The test system merely triggers the built-in test on a chip and monitors the result. There is no assurance that the built-in test is completely accurate or covers the entire function of the circuit, and there is no reliable way to automatically test the interconnecting bond wires within an integrated circuit, so some margin for error re mains. It is this margin for error that contributes to the errors that will escape detection at the in-circuit tester. Testing integrated circuits at reduced cost is covered by an IEEE Standard and was formerly called JTAG, named after the group that got this technique going, the Joint Testability Architecture Group. This technique reserves a portion of each IC as a 'test interface', and gives the test designer a way to look inside a complex integrated circuit. This method allows the same interface to push signals into internal nodes inside the integrated circuit to evaluate their effect. This 'window' into the nooks and crannies inside of an integrated circuit are used to for diagnostic purposes. In circuit testers may be unable to evaluate the health of a large and complex integrated circuits, and this will affect the accuracy and reliability of computer-aided diagnosis. Most power supplies do not have Integrated Circuits large enough to worry about this type of testing in the near future, with rare exceptions. Manufacturers of some large scale integrated circuits provide the ability for these chips to test themselves. This 'built in test' simplifies the task of the test designer, and lowers manufacturing costs. The test system merely triggers the built-in test on a chip and monitors the result. There is no assurance that the built-in test is completely accurate or covers the entire function of the circuit, and there is no reliable way to automatically test the interconnecting bond wires within an integrated circuit, so some margin for error remains. It is this margin for error that contributes to the errors that will escape detection at the in circuit tester, but will be located by Loop Testing.

10

Test Process Tools

Beyond the physical test equipment used by a manufacturer, the suite of test process tools that are available include a broad spectrum of test strategies. Competent process engineers take time to understand the advantages and disadvantages embodied in each of the strategies. The relative advantages of each process type are influenced by several variables, only a few of which are under control of the process engineer. Processes and test equipment constantly improves. New methods of analyzing information are developed regularly. Component manufacturers change their processes and designs change. To maintain the optimum type and amount of testing over the long term, test strategies must evolve as well. Obviously, some test processes are inappropriate for certain manufacturing styles or methods. Some types of test are best employed with low-volume processes, while large-volume manufacturing is best served by other test process types. Some test strategies work best when a small number of parts must be assembled, while others become more effective where there are a large number of parts in each assembly.

Available Tools

Ship to Stock "Ship to Stock" is a process where a customer strikes an agreement with one or more suppliers to provide parts that meet prespecified quality goals. This agreement goes beyond the specifications on the manufacturer's data sheets and includes the types and extent of testing performed on parts before they are shipped to the customer. The customer pays an increased cost to the supplier for this service, but the higher price is usually more than offset by the savings to be gained by eliminating or dramatically reducing the incoming inspection step for those parts.

Incoming Inspection   The earliest stage of test under the direct control of the customer is the incoming inspection step. Customers perform simple tests on parts they receive from their suppliers to determine if the parts are within specifications. Since the parts are tested before assembly, the cost of rejection is low. Some customers have contracts with their suppliers to return unaccepted parts for credit, further lowering the cost of fault detection at incoming inspection, and providing timely feedback to the vendor.

Test During Insertion   Measurement of parts during the insertion and assembly process will reduce the potential inclusion of incorrect parts in the assembly. Test during insertion is also used to check the part stream for depleted part bins, which helps ensure that assemblies have no missing parts. During assembly, the tests are limited to measurement of part value, as parts in stock are assumed to be perfect, and the class of errors to be detected are limited to wrong parts and missing parts.

In-Circuit Testing   After assembly, the completed electronic module is checked to determine if all parts are in the proper place, within acceptable value tolerance, and have been soldered firmly to the module without shorts. In-circuit testing uses sophisticated "guarding" techniques to electrically isolate each component from the surrounding circuit, so that each can be tested as if not a part of a larger assembly. Some in-circuit testers (ICTs) are capable of measuring diverse parameters of passive and active components, and are able to discern the difference between parts that may have appeared similar in tests performed during insertion.

Module Functional Testing   A module that has been assembled with all of the proper parts in the proper places and without inadvertent shorts or opens may be expected to operate properly, but the verification process involves module functional testing. A tester is used that places a completed module in an environment with controlled stimulus and load. In this way, modules that are assembled into units that contain several modules may be tested and diagnosed individually. Module functional testers may be general-purpose machines or specially designed systems, but the purpose is the same. The tester creates the stimuli to be applied to the module in the larger assembly, and measures the response of the module to those stimuli.

Unit Functional Testing The various modules of a unit are assembled and the entire unit is tested to determine if there are subtle interactions between the subassemblies that might adversely affect the operation of the unit as a whole. At this stage many of the adjustable components that ameliorate the differences between subassemblies are monitored and set. A unit that successfully completes this step generally is considered electrically sound. Unit functional testing may last a minute or two, and is performed as a gross check of the, interplay of subunits and not as a measure of reliability.

Environmentally Controlled Stress Testing (ECST) A system that is electrically sound may still be subject to long-term effects of vibration, temperature, and humidity that will affect the reliability of the device. A system for evaluating the long-term reliability of units is termed Environmentally Controlled Stress Testing (ECST), or more simply burn-in. This testing is often performed on every unit, although the details of the stresses applied and the time durations are specific to each manufacturer, and each product. Occasionally, samples of a product will be given extended stress to validate the claims of the designer about the ruggedness and reliability of the device. This extended stress testing often simulates conditions beyond the stress limits of the design in an effort to accelerate failure, and therefore induces damage in the product. This makes those units that undergo extended stress testing unsuitable for use by customers.

Final Testing Just before placing the completed unit into inventory, a complete specification check is performed on each unit. This test allows the manufacturer to determine that each unit placed into inventory has met the entire suite of specifications. If the completed units are being shipped to a customer as part of a ship-to-stock program, any tests required by the customer are performed at this point.

Ongoing Reliability Testing On a periodic basis, units are removed from completed inventory and subjected to a gauntlet of stresses and tests in an effort to determine whether these devices meet the long-term reliability goals of the product. This testing is similar to the extended stress testing mentioned above. It is performed after units have been placed into inventory, and is used to measure the effectiveness of the test processes rather than the quality of the design. The product design subjected to ongoing reliability testing must have demonstrated ability to withstand the extended stress testing that verified the design reliability goals before this testing is considered valid.

Tool Descriptions

Ship to Stock   The primary ingredient of a ship-to-stock agreement is trust, since the customer is transferring the responsibility for an important quality assurance step to someone outside of their direct control. The cost of replacing errant parts may exceed the accumulated benefits of a ship-to-stock program if the quality level of a part begins to diminish and the reduction in quality is undetected until much later in the manufacturing operation. Ship-to-stock programs are not capital intensive, but require a moderate investment in time to investigate and monitor a supplier. The supplier will usually have the requisite test equipment already in place on the production line, and the changes to the process for a viable ship-to-stock program are usually related to closer adherence to process limits or better data collection. The ship-to-stock program is a legal process for the customer, and potentially a minor process modification for the supplier. The customer, by virtue of the ship-to-stock agreement with the supplier, is purchasing preferential treatment from the supplier in the form of additional testing and cooperation.

Incoming Inspection   No manufacturing process is totally error free, and defects will eventually surface. The object of testing and inspection processes is to detect and correct defects as early in the manufacturing cycle as possible. The earliest place in the manufacturing cycle that is under the direct control of a manufacturer is the incoming inspection step.

Defects detected during incoming inspection are very inexpensive to correct. Components rejected by incoming inspection often may be returned to the supplier for credit. Where suppliers do not give credit for rejected merchandise, parts may be discarded. In either case, the cost of rejecting a component at incoming inspection is small in comparison to the cost of detecting and correcting a problem later in the manufacturing process.

Information on rejects and failure rates collected at incoming inspection is useful in three ways. The purchasing department can use the information to leverage better pricing from the supplier, and to compare the quality of multiple suppliers. The vendor is able to use the information provided to them as another means of process improvement. The design engineering group may find the failure information from incoming inspection useful as a means of predicting early failures in components. The goal of incoming inspection is to produce a stream of acceptable components for later assembly operations.

Tests at incoming are commonly a subset of the full specification test for each part. The incoming test engineer must select those tests that will most likely detect part failures and also those tests that measure parameters with the greatest impact upon the reliability of the assembly. It is meaningless to test parameters of a device that do not matter to the design. For example, there is often little reason to test the drift of a capacitor's value relative to temperature, although this parameter is specified by virtually all suppliers; in most cases this parameter is not critical to the acceptance or rejection of one capacitor over another within a shipment. In the case of capacitors, the incoming inspection test will include a test of the nominal capacitance value and possibly the leakage resistance. Some electrolytic capacitors may additionally be tested for series resistance and series inductance. Resistors are often just tested for their resistance value. Bipolar transistors are tested for DC current gain and possibly reverse breakdown, while MOSFET transistors are generally tested for transconductance. Simple integrated circuits (ICs) may be tested for a subset of their parameters with more complicated circuits have commensurably more intricate incoming inspection requirements. The tests at incoming inspection are usually very generalized for a number of reasons. In the first case, the components are assumed to be good, and this is a test to verify the supplier test process. The supplier is expected to perform the complex and complete test suite to ensure that components meet or exceed their specifications, and customers cannot be expected to double check every parameter of every component.

Remember that the cost of detecting and repairing errors in a unit climbs dramatically as the errors are detected later and later in the manufacturing process. The supplier's final test step has the lowest reject cost for an equipment manufacturer, and conversely the final test has the highest reject cost, so a supplier that defers final testing to a customer's incoming inspection may actually see economic sense in reducing his or her manufacturing cost while impacting the customer in what appears to be a small way. This delay of test from the supplier will increase the cost of the customers incoming inspection step, and the accompanying higher reject rate will do little to enhance the supplier's quality image in the customers' eye. Since few suppliers ship all of their output to a single customer, the duplication of equipment at each customer site for testing will certainly exceed the cost of the same equipment if used by the supplier. Economy of scale allows the manufacturer to more easily amortize the costs associated with the final testing of a component than any single customer. In addition, the supplier has considerable expertise in the design of the product that few customers either have or desire, and this expertise, allows the supplier to more quickly analyze the results of product testing so that the design may be improved or a process glitch may be resolved.

Test During Insertion   When properly tested components are brought into a mature assembly operation, there are two mechanisms for error: an incorrect part may be inserted in place of the correct one or a part may be omitted. By measuring parts as they are inserted during a mature process, the likelihood of either of these scenarios is slim. The term "mature process" defines a manufacturing system in which assembly of subunits has been proven to be performed correctly. Testing during insertion may be performed manually by human assemblers, but is best done by automatic insertion equipment. Human control of testing is prone to an assortment of distortions and inconsistencies that are not found in machine directed testing. The test performed at the point of insertion is very inexact, and is often much less demanding that the tests performed at incoming inspection (which themselves are a small subset of the full test for conformance to specifications). The intent of testing during insertion is to detect conditions that would lead to:

- Components incorrectly oriented such as diodes, transistors, ICs
- Components of the wrong value, primarily resistors and capacitors
- Components of the wrong type such as a diode in place of a resistor
- Components not inserted at all, or missing.

At the completion of the insertion step, the unit should have good parts inserted at the proper place and in the proper orientation. There will be no missing parts and all parts are very likely to be the proper parts for each position. There are considerable opportunities for inappropriate parts to be placed into an assembly in the presence of test during insertion. Testing during insertion can be subverted by subtle differences in components that have critical effects during device operation. A design may call for a transistor with a particular gain-bandwidth product. This parameter is too complicated to measure during insertion, requiring a more controlled environment than is found within the insertion mechanism, and thus a transistor with less bandwidth may pass the insertion test without difficulty and yet fail (perhaps with dramatic flair) when the assembly is operated. Other parameters that can escape detection at insertion include logic functions of ICs, power dissipation of high-energy devices, breakdown voltage of semiconductors, and leakage current of active or passive devices. Where components are subject to these confusions of identity, testing of the components is warranted at a later manufacturing step to verify the identity of those parts. Some insertion testing is based upon automated visual inspection. A television camera is connected to a specialized computer that searches the image and compares what the camera sees to a stored pattern or template. A component that is badly inserted or missing will be detected by these vision systems, although they are even more susceptible to errors of identity than systems that perform electrical test during insertion. After the insertion step, any manual assembly operations are performed. Manual assembly is often required where components are too massive for automatic placement or where mechanical fasteners like screws, staples or rivets are used. Manual assembly steps must employ manual component testing to produce the same quality as automatic insertion systems provide.

In-Circuit Testing An electronic assembly, usually in the form of a circuit board or module, may be tested after component insertion and soldering. The initial test of a completely assembled module is often performed by in-circuit test systems. In-circuit test systems include a mechanism for electrically monitoring many points within the module simultaneously. This multiple electrical connection is often accomplished by a multitude of spring-loaded metal pins that protrude from an insulating panel. These metal pins, called test probes individually or a bed-of-nails collectively, are each connected to an electronic switch matrix that can be directed by the tester control computer to route the signals from the test-probes into specialized instruments within the tester. The switch matrix can make hundreds of connections and disconnections every minute. The job of the tester is to electrically test each component within the assembled product by electrically isolating each component and testing as though the component were not part of a larger unit. The technique for electrically isolating each component is called guarding. Guarding uses a simple fact of electricity, if the voltage on both ends of a component or wire is identical, then the component or wire may be removed from the circuit, or may be treated as if it did not exist. See Figure 10-1.

There are several sources that detail In-Circuit Testing basics very well. Three of them are:

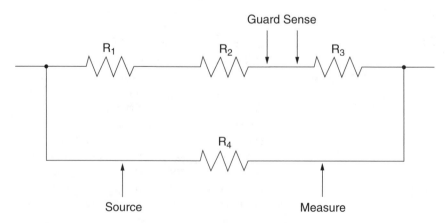

Figure 10-1. Circuit Guarding

- ———, *Introduction to In-Circuit Testing*, GenRad, Inc., Concord, Ma, 01742
- Graig T. Pynn, *The Low-Cost Board Test Handbook*, Zhentel Inc, Walnut Creek, Ca 94598
- ———, *The Primer of High-Performance In-Circuit Testing*, FACTRON Schlumberger, Latham, NY, 12110

Different testers have different capabilities. There are several types of testers classed by the type of testing that they can do. They include:

- 2-Wire
- 3-Wire
- 4-Wire
- 6-Wire
- 8 & 10 Wire

2-Wire testers are the simplest and can only measure a few components in the circuit. **3-Wire** testers have the ability to add a guard wire to keep currents from other components from blurring the measurements. 3-Wire testers work quite well, except where the component values are not near the center of the range. However, low or very high values are far less accurate. The **4-Wire** ICT has the ability to make measurements for power supplies components. It has the measurement capability to measure chokes of a few microhenries, resistors to less than an ohm, and large electrolytic capacitors. **6-Wire** ICTs have added wires, but no added capability. Some believe that the added wires add extra noise to the measurements being made. The **4-wire** system can duplicate even the six-wire measurements by doing multiple tests, and yield greater accuracy and less noise. **8 & 10 Wire** In-Circuit Testers are used by the National Institute of Standards and Technology (NIST) and others requiring extreme accuracy at an extreme price. NIST was formerly the National Bureau of Standards. These systems are never used in manufacturing systems.

It is the authors opinion that In-Circuit Testers similar to the **GenRad GR2280i** 4-wire In-Circuit Test System are the best systems suited for power supply ICT and Loop testing. This opinion is based upon both the capability of the 4-wire system, the auto test generation capability and ease of programming of the system

Module Functional Testing

Where an electronic assembly is a portion of a larger electronic unit, many manufacturers test each subassembly on a tester that both provides the signals each assembly needs and measures the signals the unit produces. Functional testing is usually done in a piecemeal fashion, testing each of the subfunctions within the unit one at a time. This disection into smaller simpler tests serves two purposes: it allows a direct correlation between the results of each test and the portion of the assembly where the fault is located, and reduces the need for more than a handful of stimulus sources, loads, and measurement instruments. The assemblies that fail functional testing often are legitimately faulty, although the converse is less accurate; passing functional testing does not assure fault-free assembly. Functional testers include a number of stimulus sources, including power supplies, function generators, and logic signal sources. Measurement units in these testers are able to measure voltage, current, and resistance. More complex testers are able to determine frequency, pulse duration, and other waveform phenomena. The interconnection switching scheme for these testers is usually much less extensive than the switch matrices found in ICTs. What functional testers lack in number of switching channels is offset by the bandwidth of the switching system. ICTs require very low bandwidth of their switches, because much of the testing is done at DC or frequencies below 1 MHz. Functional testers, on the other hand, may generate, route, or measure signals that range into the hundreds of megahertz. The signal integrity concerns at these frequencies are more complex and the designs of the functional tester switch matrices reflect this concern. To compensate for the relatively limited signal switching apparatus, much of the signal routing hardware is part of an adapter system that is placed between the tester proper and the device to be tested. Power supply testing presents unique problems in comparison to most circuits, because of the presence of a feedback stabilized control loop. An error injected into the loop by a faulty component will propagate around the circuit loop and appear everywhere. The control loop may actually stabilize in the erroneous condition, making analysis of the failure difficult or impossible. The common method for isolating failures in the control loop is to break the loop and treat the circuit as an amplifier. Since errors cannot propagate beyond the point where the loop is broken, the failure is usually located at the point where the loop-cum-amplifier deviates from expected operation. Figure 10-2 shows a simple power supply control loop and Figure 10-3 shows the loop configured as an amplifier in a typical application of loop testing.

Unit Functional Testing

Modules are assembled into larger units, and the variations from one module of a type to another of the same type may precipitate failures when included with other modules. Additionally, adjustable components that are used to

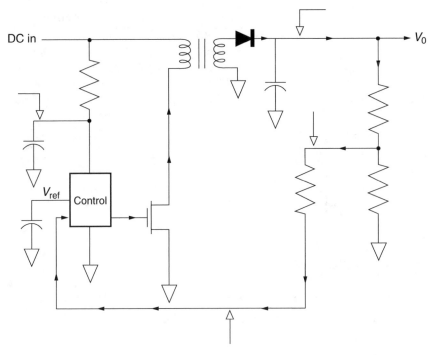

Figure 10-2. Simple Power Supply Control Loop

modify signals that travel between modules are adjusted during unit testing. Much of the unit test is designed to detect interaction faults between modules, and errors introduced by the assembly of modules into the larger unit. Unit testing is often the first point in the test where the specifications of the completed product are verified. Very mature high-volume processes simply test that the modules are interconnected properly and that the unit operates. Since the unit generally requires a power source and limited loading and measuring capability, the unit tester is almost always less complex than the system employed for module testing. In the cases where the assembled unit has specifications that cannot be determined at the module level, some additional equipment may be needed at unit test that is not required at module test. One case where the unit test is more complex is on very high voltage power supplies, where the step-up transformer may be mounted to the chassis separately from the modules, and thus the specifications for output voltage and current depend heavily upon the interplay of the control module and the transformer which first encounter each other at unit test. Unit functional testing may require very-high-energy variable power sources to simulate the conditions of line variations for an assortment of environ-

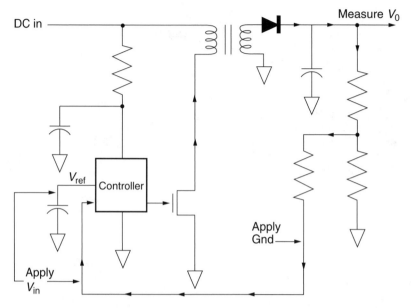

Figure 10-3. Power Supply Control Loop as an Amplifier

ments, particularly where the product is shipped into the international market. In general, the loading and measuring equipment will not change from a domestic to an international product.

An adjunct to unit functional testing is HIPOT testing, which verifies the insulation between the input line and the chassis ground. HIPOT testers are usually separate devices rather than being part of the unit functional tester, due to the specialized high-voltage circuitry and time required to perform HIPOT testing that would severely limit unit tester throughput. During HIPOT testing, several thousand volts are applied between the line and chassis ground. The HIPOT tester measures the amount of current that leaks between the line and the chassis ground, and where that current exceeds a predetermined limit, a failure is declared. HIPOT testing generally follows a sequence of varying voltages and durations. It may be a series of three on-off cycles each lasting 4 min on then 30 S off with voltages that range up to 1500 V.

Environmentally Controlled Stress Testing (ECST)

Units that have passed unit testing have proven that they electrically meet specifications in the absence of unusual stress from temperature, vibration, or humidity. The commonly accepted failure likelihood curve for a product is a bathtub-shaped curve. The first part of the curve is a steeply declining possibility for failure over time which represents so-called infant mortality. The bottom of the curve that slopes gently upward over time depicts how likely a failure is during the operational life of a product. After some time, the curve begins to trend upward indicating the product is more likely to fail due to wearout. This curve is called the bathtub curve due to its shape. The three regions of the curve are termed early failure, normal life, and wearout. It should be recognized that the curve is based on statistical likelihood, and is not valid when applied to small populations of devices. What this means is that it is impossible to predict when a particular device will fail under normal use, but it is possible to characterize the reliability of an average unit within a large group. A particular device may fail immediately when it is first used, while another device may last dozens of times longer than the statistical evidence would indicate. Using statistical models and methods, an engineer may evaluate the likelihood of a failure of an average unit of a product. ECST or burn-in as some call it, is employed to exacerbate the failure mechanisms that contribute to infant mortality failures. Many of these mechanisms are related to mechanical problems within the parts or between the assemblies of a unit. Remember, the parts have been shown to be electrically operational by unit testing, and we assume that the product is using a proven mature design, so that extraordinary electrical stresses are unlikely. The units that pass ECST may be assumed to have reached the normal life portion of the failure likelihood curve and thus have a very high potential for survival. Use of ECST consumes a block of time that represents the early failure portion of the reliability curve, and the applied stress acts to more quickly cover that portion of the curve, essentially aggravating many failures that would normally appear early in the useful life of the product. ECST does not eliminate early failures but simply encourages them to happen before the units are shipped to a customer. Stress during ECST comes in many forms, but the most common form is heat. In the normal use of a product, heat is the most likely stress, so test designers often use heat as the stress of choice. Since many power products produce copious heat as a by-product of their normal operation, it is a simple matter to confine the units being tested so that their excess heat collects around them as they operate. This is simple, inexpensive, and remarkably effective. Heat causes materials to expand, and the rate at which materials expand varies widely. Two parts that share a mechanical interface like a solder joint, a weld seam, an adhesive bond or mechanical fastener will loosen as the expansion stresses wear the interface. This effect can be intensified by repetitive heating and cooling of the interface. In normal operation, the unit might accumulate enough heat cool cycles in a month to precipitate a failure, but

ECST can be used to simulate those thermal effects in less than one day. Many electronic components fail from application of heat rather than thermal cycling. Capacitors can dry out, hot spots can develop within resistors, and transistors may dramatically change their behavior in the presence of heat. A bipolar transistor will have an improved ratio of output current to input current (gain) as temperature increases. As the device increases its gain, it is prone to produce more heat, which further increases the gain, and the situation feed upon itself until the device destroys itself. Even if a device is not prone to thermal runaway, the heat within a device must not exceed certain thermal limits. Designers use heatsinks to help remove heat from devices that either are very sensitive to heat damage or produce enough heat within themselves to cause the damage. Within a semiconductor device, the junction temperature (or Tj) is not supposed to exceed 200C or the device will be damaged. It is unimportant whether the device produces the heat itself, or whether the heat comes from sources external to the device, but 200C is the limit. Many engineers use lower limits to decrease the possibility of early failure. The semiconductor junction which may be damaged is imperfectly linked to the ambient air around the device, and thus the air temperature, or the device case temperature is an imperfect measure of the semiconductor temperature deep within the device. Manufacturers of semiconductor devices specify the maximum amount of temperature difference between the case of the device and the junction in relation to the applied power. The so-called thermal resistance is specified as °C per watt. If a device has a thermal resistance from the junction to the case (T jc) of 2.5 °C/W, and the power we wish to apply to the device is 30 W, then we must design the cooling system to assure that the device case never exceeds $200-(30 * 2.5) = 125$ °C. Heat sink manufacturers and thermal compound manufacturers specify the thermal resistance of their products as well, so the sum of the resistances must be taken into account. For example, assume we use a thermal compound with 0.15 °C/W thermal resistance and a heat sink with 3.1 °C/W thermal resistance with the device mentioned above. The air temperature passing over the heatsink must never exceed $200-[30 * (2.5 + 0.15 + 3.1)] = 200-(30 * 5.75) = 27.5$ °C (or about 72°F) to avoid failure of the device. It can be seen that a small increase in the thermal resistance can spell disaster for devices that must be temperature controlled. ECST will place demands upon the unit cooling system by lowering the margin for thermal resistance tolerance. Where thermal resistances accumulate beyond design limits, devices will fail under ECST. Some reliability engineers will thermally shock the systems to further accelerate failure mechanisms. This involves artificially cooling nonpowered units to a temperature well below room temperature (but still within the design limits), and then applying power, so that the components and connections rapidly gain heat.

Sometimes, systems will be operated until they reach a temperature near the top of their design limits, and as soon as power is removed, artificial cooling will be applied to quickly remove heat from the system. Each of these techniques is used to aggravate physical expansion and contraction phenomena. A burn-in process may be as simple as operating each unit into a resistive load for 72 hr in the factory ambient environment. More complex ECST will specify temperature cycling and power cycling requirements, and may go as far as specifying a varying load cycle. The ECST requirements may be set by company reliability policy, or may depend upon special design considerations. Each manufacturer develops an ECST strategy appropriate to the product.

Final Testing

Once a unit has completed the gauntlet of stress tests, it is subjected to one final test process before being placed in completed inventory. This test verifies that the specifications of the device are within acceptable limits, and that all functions operate properly. This test is very similar to the acceptance test that a customer might perform. By working with customers to tailor this test to the customer specifications, a ship to stock contract may be struck. In this way, the customer will have the critical acceptance measurements performed with no capital investment, and only a small increase (or no increase) in the product cost. To effectively use final test as part of a ship-to-stock program, it is vital to have certification of the final test process. Periodically, known good and known faulty parts should be passed through the final test to determine if the process can still separate the wheat from the chaff. Accurate record keeping which logs the date and time each unit is tested helps to locate units which deserve to be retested should the test process fail a certification operation. Certification may be triggered more often than the schedule would dictate by abnormally high or abnormally low numbers of failures, or the absence of failures. Again, record keeping allows this abnormal condition to be identified and suspect parts to be located for retesting. Final testing is generally used as a process gate and not as a diagnostic operation, so failures detected here are returned to an earlier test process step for diagnosis and correction. Final testing may be performed by persons whose primary function is inspection rather than electrical test, as diagnosis is deferred to another process step, and in many factories final test is performed by the same people who inspect units for missing controls, damaged cases and connectors, and other cosmetic defects. Units that pass final test (and cosmetic inspection) are placed in cartons and sent to inventory. In the case of ship-to-stock, the units are packed and sent to the customer.

Ongoing Reliability Testing

Some forms of product testing last well beyond the limits applied to the day-to-day product. Reliability engineers will often remove several units from inventory, or divert a few units exiting from final test, for a special form of testing called ongoing reliability testing. This testing is very similar to the ECST mentioned above, except that the stresses may be more tortuous and the duration considerably longer than the usual ECST test. The goal of this testing is to determine the useful life of the product, and find any missed early failures by using ECST which only attempts to induce very early failures. By accumulated experience, or some accepted acceleration formula, the reliability engineer determines the level and duration of stress that will simulate a product lifetime of several years within the span of a few weeks. The product is subjected to the test and its specifications monitored either continually or on a periodic basis during the test. Unit survival times are recorded and extrapolated to determine the product's expected normal life. The projected lifetimes of several units are analyzed by statistical methods to determine if the product meets the mean time between failures (MTBF) goal. Should the analysis prove that the projected lifetime of the product is shorter than the reliability goal of the designer, either a design change or improved product testing may be required. Some forms of ongoing reliability testing continue until all units fail under stress, while other forms set a maximum time limit to the test. Where a time limits is imposed, the duration is selected such that a unit that lasts the entire time is considered to have a projected lifetime of at least twice the MTBF goal. Units with projected lifetimes greater than twice the MTBF goal will contribute little to the proof of reliability. By testing until all units have failed, it is possible to evaluate the MTBF goal and determine if it may be extended without risk. If the statistically projected lifetime of a product is much longer than the design goal, the reliability engineer may determine that the acceleration factor applied to the product under stress is inappropriate, and that the stress must be increased or that the ratio of real time to simulated time must be modified. Careful analysis of failures can aid the designer in correcting the present design or safeguarding future designs. The failures detected under life testing are usually very different than those found during manufacturing tests, unless the components with the highest reject rate also contribute significantly to the long-term failures. The manufacturing process can use a handful of these test processes, or (where appropriate) all of them. Many of these processes are simply not useful in small-volume situations, particularly those test steps that rely heavily upon statistical operations on large populations of units or that require specialized machinery. It is the job of the test engineer to evaluate the cost of each test process step against the cost of not performing that particular test, and determine by the relative costs whether or not the process step makes economic sense. In the final analysis, a manufacturer makes a product for the express purpose of economic profit, so all options should be examined in the light of long-term economic benefit.

11

Choosing Testers

The selection of test equipment is a complicated and inexact process. Specific requirements of the manufacturing process and fiscal restraints are but two of the criteria that must be considered when choosing testers. Equipment choices are particularly difficult in an immature product situation, where the projected volumes and test requirements are not as firm. The equipment strategy is very different in a low-volume uncontrolled situation than it is in a stringent high-volume manufacturing scenario. In an uncontrolled manufacturing operation, the unit specifications may be fluid or the design goals not expressly stated in an effort to be most responsive to customer needs. This often blurs many of the specifications for testing. A conscientious test engineer will err on the side of conservatism, choosing equipment that has the ability to cover a wider range of signals as well as measuring to a tighter tolerance than the manufacturing process may demand. This conservatism allows the test engineer to have confidence in the test results.

Confidence

Testing is simply a confidence reinforcement. In its most basic form, a test is the comparison of an entity with unknown characteristics against a reference. This effort is required to analyze the difference in characteristics between the reference and the unknown. This comparison is meaningless if the reference is not predictable or if the difference analysis is unstable or misunderstood.

Predictability

Without predictability there is no reason to perform test at all. Comparison to a reference that is changing loses significance as the degree of change increases. A test engineer understands these facts and selects the equipment that will have the appropriate degree of predictability so that the manufacturer may have sufficient confidence in the quality of the product to make reliability predictions and develop warranty parameters. In the face of imprecise product specifications, the test engineer must exaggerate the predictability requirements of test equipment to maintain the required confidence. When in doubt, test engineers buy the best equipment they can afford. Uncertain product specifications drive the cost of test equipment purchases up as test engineers attempt to keep quality under control. This uncertainty can also increase the costs of equipment ownership in the form of more complex or more frequent operator training and more frequent equipment calibration or repair. Additional costs can be incurred where unforeseen product enhancements demand extra instruments or extra features in existing instruments. The most cost effective test equipment purchases are made when the product requirements are well understood. Planning is vitally important. The person charged with selecting a tester or several testers must examine a wide range of process and product parameters before selecting the appropriate solution.

Product Volume

The number of units to be produced will affect the degree of automation in the equipment to be selected. Automation requires considerable engineering before the equipment is used in manufacturing. Simple automation allows the operator to recall a few prerecorded instrument configurations, which will reduce setup errors and save time in situations where instrument controls are routinely adjusted several times during test. The test engineer will develop the sequence of configurations and the commands necessary to produce the desired result, and store these commands either within the instruments or on some medium that may be read by the instruments. In higher-volume applications, the entire test may be under the control of a test computer. In this instance, the program executed by the computer must be created and tested before the tester is used. This programming will add cost to the test process, but may be amortized over a large volume so that the cost per unit tested is only a few cents. The cost of automated equipment is higher than semiautomatic or manual equipment, but the use of automated equipment allows much faster testing thus reducing the cost per unit tested. A complex situation arises where a product is planned as a low-volume unit and sudden customer demand catapults the product into a high-volume arena with little or no warning. The testers, processes, and training used for low volume processes are usually inappropriate for high-volume manufacturing, and the long-range planning that a high-volume process demands cannot be completed without affecting shipments. A viable strategy is to develop test plans for low-, medium-, and high-volume manufacturing of each unit; selecting test equipment, outlining training requirements, and developing a strategy for migrating from a low-volume sphere to high-volume manufacturing. Some companies use high volume techniques and testers for medium- and low-volume products because they have the goal for each product to eventually require high-volume shipments. These manufacturers achieve economic benefit by sharing the expensive automated testers among several products. They purchase enough tester capability to handle expected plant output until some point in the future. The amount of time that would be required to order and receive additional testers is usually the amount of time into the future that these manufacturers plan. In this way, should product demand surge, the existing test programs would be duplicated onto the additional machines as they arrive with no additional delays.

Long-Term and Short-Term Strategies

Short-term planning carries less risk and requires much less analysis than does long-term planning. However, short-term planning requires constant review which a solid long-term plan avoids. Short-term planning is appropriate where the product manufacturing cycles are short; that is where a product is manufactured under a specific finite contract, or where the entire output of the manufacturing process is delivered to a single customer. In this scenario, the customer (or contract partner) determines the demand and purchases a limited quantity of a product with sufficient advance notice of expected volumes to allow extremely precise knowledge of the manufacturing capabilities and test equipment capabilities. The customer assumes responsibility for all long-range planning, and where errors of demand prediction occur, the customer may renegotiate the purchase agreement or even seek additional suppliers. Where the product is sold on the commodity market, or is manufactured as a standard product available for purchase by several customers, the onus for long-range planning reverts to the manufacturer. A manufacturer attempts to maintain a certain amount of inventory 'float' stock, which acts as a shock absorber when demand fluctuates or may create a manufacturing process capable of fast response time, where required. The number of units in inventory is based upon the difference between normal and abnormal demand and the ability of the manufacturer to respond to the changes in demand. The faster a manufacturer is able to respond to increased or decreased demand, the less float stock is required. Conversely, the slower a manufacturer's response, the greater the number of units required to cushion the changing demand. The manufacturer's ability to change the process output volume is measured in relation to the stated shipment delay (delay from receipt of order). The number of units in float is determined by the relative speed of response and the difference between normal and maximum shipments. If the manufacturer has the ability to affect process output volume rapidly, little or no float stock is required, but if the manufacturer needs time beyond the shipment delay to modify the process output volume, the amount of float stock may equal the amount shipped each month. Product volumes may not be static, and this is often the case on new products or where external factors may affect demand. Product demand often follows a curve similar to that shown in Figure 11-1. The demand builds over a period of time and eventually peaks, then gradually tapers off.

Competition can modify the shape or slope of this curve and thus affect demand for a particular manufacturer's product. Demand is the primary component of any long term strategy. Product trends impact the decisions made in long-range planning as well. These trends are very specific to the intended market, but may involve tightening specifications, increased reliability, or a trend to higher-power density or more functions. Ignoring product

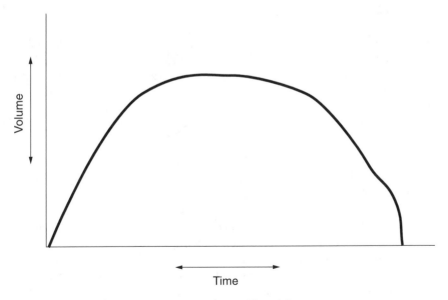

Figure 11-1. Product Life Cycle Curve

trends may lead to selection of testers that will be ill equipped to satisfy the demands of the process when newer designs are made.

Short-term planning looks only at the current needs, and may be sufficient where the trends do not indicate products that are significantly different than those made today. Obsolescence of a product also drives long-term strategy. There is little economic sense in choosing a tester for its ability to address the unusual trait of a product that may be manufactured for a brief period. The long-term view would dictate that a more general test strategy be developed, and the unusual characteristic of the unit be tested by other means. There is another consideration when choosing testers which few manufacturers take into account: what will be done when a tester is no longer appropriate to the process? Specialized equipment that demands a premium purchase price may not be useful throughout its tax depreciation cycle, and may have little or no value on the used equipment market. Generalized equipment has a greater potential for application to other test processes in the same plant or to lower volume products, and has a higher resale value on the used equipment market. Use of adapter electronics and connectors to allow generalized equipment to make unusual measurements or serve in an unusual capacity will improve the likelihood that the purchase will make economic sense. In the balance of this chapter we will examine the characteristics that will have the greatest impact upon the process of choosing test equipment.

Capital

The essential goal of any business is to generate profit. Profit is the difference between the cost of manufacturing a unit and the price for which the unit is sold. The cost of producing a unit is divided into capital and components. The components are under control of the design engineer, but capital is the province of the process designers which include test engineers and manufacturing engineers. The primary components of capital expenses are:

- Process
- Plant
- Personnel

Process is the equipment used to actually assemble the unit, and includes items like tools, conveyors, soldering systems, and the various carts, bins, and furniture used in a manufacturing environment. Test includes the instruments, in-circuit test systems, adapter systems, test control systems, and any environmental stress units such as ovens, vibration tables, humidity chambers, and burn-in racks. The plant encompasses the utilities such as power, lighting, water, the physical building, and optionally pressurized air, process vacuum, steam, and process chemicals. The last element, personnel, represents the human element required to assemble, test, inspect, diagnose, repair, package, and transport the manufactured units. It should be noted that capital does not include overhead or indirect expenses, such as the engineering, administration, or sales costs associated with a manufacturing environment.

Evaluating Specifications

The differences in specifications between testers comprise the most basic means of comparing one system against another. Manufacturers, profoundly aware of the reliance that engineers place in published specifications, often resort to a practice known as "specsmanship," where the published specifications are presented in a less than objective manner. There are several manifestations of this practice, ranging from the relatively benign custom of placing the most impressive specifications at the top of a specification sheet to the intentional misstatement of capabilities. An accuracy specification may be reported without indicating the temperature range over which the accuracy is maintained, or a precision specification may not be valid on all ranges of an instrument within the tester. A comparison between testers must be conducted on equal terms. Manufacturers seldom equivocate when asked directly about the specifications under specific conditions. The responses from such inquiries are a fair place to begin a comparison. Among the specifications common to most testers and instruments are:

- Accuracy
- Resolution (or precision)
- Traceability (and calibration)
- Impedance matching
- Reference standards
- Mean time between failures

Accuracy is a the measure of the difference between the actual value of a parameter of interest (see the discussion of standards which follow) and the measurement determined by the instrument. The difference is reported as a percentage equation: [(actual measured)/(actual)] * (100.0). On most instruments, the accuracy is stated for every function and range. Complex testers may have dozens of measurement, stimulus and load functions, with several ranges for each function and an accompanying accuracy specification.

Accuracy is the quality that determines the confidence an engineer will have in a particular measurement in a particular range.

Resolution is another name for precision, which is simply the number of significant digits in a measurement or control value. To contrast precision with accuracy, contrast two imaginary voltmeters with an accuracy of 0.I % measuring a source known to produce an output of 102.55 V. A voltmeter with two and a half digits of resolution would read 103 V, while a meter with four and a half digits or resolution might read 102.47 V. Both meters would be operating within their accuracy specification, but a human examining the output of these devices would conclude that one must have worse accuracy than the other as one reads 103 V while the other is closer to 102 V. The example is fictional, for few meters would be manufactured like the imaginary four-and-a-half digit unit described above, having a resolution an order of magnitude larger than the resolution. With an accuracy Of 0.1 %, only three and a half digits can be significant; any more digits at that accuracy are superfluous. Some instrument specification sheets trumpet the resolution and de-emphasize the accuracy to disguise this very fact.

Traceability is documentation or proof that an instrument has been either directly or indirectly compared (via one or more intermediate instruments) with the National Institute of Standards and Technology (NIST) standards. Instruments do not remain accurate forever. During manufacture and periodically thereafter, an instrument must be calibrated, which simply means to adjust the instrument to operate as accurately as possible. In the United States, the NIST develops and preserves 'standards.' Many of these standards are physical devices that produce electrical signals which are the considered to be perfectly accurate to a known precision. Instruments which are traceable are generally considered to be more accurate than instruments whose accuracy is not traceable to NIST. Duplicates of the NIST standards are available and may be compared to the official standards. The difference between the two standards are recorded and this record accompanies the secondary standard as a means of correcting measurements taken from it. These secondary standards and tertiary standards derived from the secondary standards are termed reference standards. These points of reference are used to accomplish traceability. Reference standards allow the tester or instrument to be periodically adjusted to a known accuracy. On sensitive instruments, the resolution may be limited by assorted signal distortions collectively called noise. In the presence of extraneous signals, instruments with otherwise impeccable resolution and accuracy specifications must be operated with derated specifications. Careful attention to signal integrity issues within a test system allows instruments to be used without derating their specifications.

Impedance matching is a signal integrity concern that is receiving increasing attention today. The amount of energy transferred from one point in a circuit to another is directly related to the difference in impedance between the source and an interconnection wire and the difference in impedance between the interconnection wire and the destination. If an interconnection has an impedance identical to the source and destination (and by inference the source and destination impedance are identical) the maximum amount of energy will be transferred. It is virtually impossible to devise a tester that exactly matches the impedance between each instrument and the various points within the circuit being tested. A perfect impedance match is not required, but it does make testing of signals with high frequencies or fast changing edges much easier. Impedance matching allows signals with lower amplitudes to be measured with much greater accuracy on any instrument. At frequencies above a few hundred megahertz, or where signal changes happen in less than a dozen nanoseconds, another problem begins to affect an instrument's ability to analyze a signal. The speed of light and of electrical signals is roughly one foot per nanosecond in a vacuum, but becomes slower when these signals travel in other media such printed circuit boards [**the signal actually travels on the surface of the volume and materials around a wire at extremely high frequencies rather than in the wire itself**]. When a signal traveling down a wire encounters an impedance mismatch, part of the signal will pass to the destination and the rest of the signal will reflect back toward the source. These reflections affect the shape of the signal edge as measured anywhere on the wire, and in the case of high frequency signals will cause the amplitude of the signal seen at various places on the wire to vary greatly.

Reference Standards are those standards whose calibration is directly traceable to, or calibrated by NIST, with verifiable documentation.

Mean Time Between Failure (MTBF) is the expected or calculated time between failures of a system, subsystem, or component part. MIL STD 217 is one of the most often references in calculating MTBF figures for components, assemblies, and systems. When the MTBF is known, the correct cost of ownership of a product can be calculated figuring in cost and frequency of repair, number of spares, maintenance personnel, and other true life cost factors.

Types of Testers

Component Testers

Incoming inspection processes and assembly processes use component testers. Individual passive components are easier to test before connection to other arts than they are after assembly, but some active components, particularly complex ICs are more easily tested in their intended environment.

Component testers come in a variety of forms. The simplest units are nothing more than ohmmeters, LRC bridges, and transistor curve tracers. These machines are controlled and fed by human operators. More complex testers have the ability to configure themselves for one type of test or a different range or limits on command from an operator, which frees the human to attach and remove components from the interface rather than adjusting the controls of the test instruments. The most elaborate machines are completely automatic, and may accept loose components which are individually selected from the input hopper, positioned automatically to the test interface, tested, and placed in the output hopper. Some machines accept components on tape and reels. A few machines accept loose components and place the acceptable components onto tape and reels in sequences suitable for application to automatic insertion machines. Most of the completely automatic machines are found in extremely high volume operations, as low-volume manufacturing is characterized by manual processes in all but a handful of cases, and low-volume manual manufacturing cannot justify the economies of scale provided by automatic testers.

In-Circuit Testers

Testers designed to measure components that are parts of a completed circuit are more expensive than testers for unconnected components. The added cost comes from the need to provide a variety of instruments within one enclosure as well as a sophisticated signal switching system. In-circuit testers are controlled by a computer which is not found in most component testers, and the cost of producing and maintaining the control program software contributes to the increased cost of in-circuit test machines. These testers are distinguished from one another by the complexity of the switching system, the number and quality of the instruments, and the features of the system software. As the number of signals required to perform a single test increases, the size and cost of the signal switcher increases as well. The increased complexity of switching is offset by the ability to measure a broader range of component parameters to a higher accuracy than is available on testers that use fewer signals per test.

Functional Board Testers

Neophytes often believe testing is limited to measuring the performance of a completed module, and seasoned engineers place considerable emphasis upon full functional testing, believing that this test proves that the board actually operates and that other testing is superfluous. It is true that module functional test will usually detect virtually all of the faults that would be detected by component testers or ICTs, but the cost of detecting and repairing faults at functional test is many times the cost associated with detecting and correcting the same fault at earlier stages. The costs of fault detection at functional test may be reduced considerably by judicious investment in test operations at incoming inspection and insertion processes and by use of loop testing on in-circuit testers before functional testing.

Dedicated Module Testers

In traditional operations, module functional test was accomplished by using a completed unit known to operate correctly and substituting the suspect module for its twin in the unit. Failure of the unit or poor operation could be attributed to the suspect module and diagnosed with the knowledge that the fault would be located on that module. Some benchtop units merely simulated the other modules in a unit, particularly where a module required characterization over a broad range of conditions and stimuli. The cost of the physical benchtop tester was small when measured against the costs of a general purpose functional tester. The general-purpose tester makes economic sense where the number of dissimilar designs causes a proliferation of benchtop testers. Design, maintenance, operation, and storage of several infrequently used dedicated testers will eventually cost more than one general purpose tester used more or less continuously. The tradeoff between testers depends upon the number of designs manufactured by a plant and the volume of each manufactured. Onetime costs of general-purpose functional testers include fixturing, software creation, debug, and training. These costs must be factored into the analysis of test expenses along with repair, equipment purchase, operating costs, and design engineering.

Benchtop Dedicated Unit Testers

A group of tested modules assembled into a completed unit, may be tested on dedicated tester. Depending upon the amount of power being converted by a particular design, this tester may be on a benchtop or occupy a corner of the manufacturing floor. Unit testing may be a simple resistive load station with ample line power. Rigorous unit testing for variable power supplies will require additional equipment including a source of variable AC voltage, adjustable loads, and instruments to monitor unit performance. The instruments commonly found in dedicated unit testers are voltmeters, ammeters and an oscilloscope.

Combination Functional and In-Circuit Testers

A class of machines has been developed that combines the ability to perform in-circuit testing and functional testing on modules. Most of these machines are not designed to handle the power levels required and produced by entire units. These machines may be employed to validate the correctness of components using an in-circuit test, and then test hat the function performed by those components operates correctly. The cost of purchasing an in-circuit tester and a separate functional tester may be reduced by using a machine which combines both types of test, as the volume increases, the throughput of one combination tester will not sufficiently replace two specialized testers, and a decision based upon both costs and utilization will then take precedence.

Functional Testers

A number of general purpose functional testers exist, each with unique qualities and specifications. These machines include the necessary stimulus, load, and measurement instruments to verify the operation of power modules and units. Automated functional testers include signal switching systems that are less extensive than the switching systems found in in-circuit systems, and manual functional testers use dedicated harnesses and fixtures to accomplish interconnection of the unit and the testers A computer is usually employed in automated systems, while a human operator controls manual systems.

Functional Burn-in Systems

Environmentally Controlled Stress Testing (ECST) may be combined with a simple functional test to simultaneously stress and evaluate the operation of a power supply unit. The functional test used during burn-in is not as exhaustive as the functional test one might find in a functional test without stress, as the goal is to detect gross failures rather than uncover small variations in operation that do not exceed specifications. Functional test during burn-in is employed more as a pass/fail test than as a diagnostic. Manual systems may be as simple as voltmeters across the output terminals of the unit. Variable transformers may be employed to adjust the source voltage to higher and lower than normal during ECST, and loads may be dynamically adjusted during the test by the operator or simple timing circuits.

Automated Functional Burn-In Systems

Very high volume ECST requires a system with more automation than most burn-in sequences. These systems are controlled by timers or computers to produce a predetermined sequence of source voltages, loads and environmental conditions (heat, humidity, vibration, etc.). Automated systems produce more consistent stress and monitoring than is possible to achieve with human intervention The consistency of stress and monitoring removes an additional source of uncertainty in the analysis of ECST results.

Evaluation of Testers

When one must choose appropriate test equipment for a process, technical and business requirements form the basis for any decision. The technical specifications of the process will provide a set of baseline characteristics that describe the qualities of an acceptable machine. Business issues will influence the selection process toward a system with either improved or reduced specifications. In cases where a manufacturing plant already contains several different testers, the evaluation may examine the capabilities and capacities of existing testers to select the most appropriate machine for each process step. Rare exceptions to this decision making process occur where a manufacturer opts to modify a new design to avoid acquiring new test equipment.

Technical Considerations

The required capabilities of a tester are determined by the product designer and the test engineer. A list of proposed tests is developed from the design specifications, and the attributes of the tests (tolerance, precision, range, etc.) are determined. The test requirements are compared to the specifications of testers and acceptable machines are noted. Test requirements for a unit are often a transliteration of the product specification sheet, while the tests for submodules internal to the unit are based upon the characteristics of the modules as understood by the designer. Electrical functions distributed between modules complicate analysis of signals that will be measured and supplied by a module tester. Experienced test engineers do not attempt to second-guess the designer, and work with the design and product engineers to develop realistic test requirements. A test requirements analysis summary sheet is usually created by test engineering. This report details the number and types of tests that will ensure that a particular device is acceptable. The summary makes no attempt to direct the sequence of tests, although sequencing may be summarized where it is crucial to the effective test of the device (i.e., test output over voltage protection (OVP) circuits before applying input power to the unit). The tolerances and limits specified for each test provide a point of reference for the test designer so that an appropriate test system may be selected. A test engineer uses the summary to determine the stimulus and measurement instruments that will be required, then compares tester specifications to determine if a particular tester meets the requirements. In addition to the electrical specifications, the time element is examined: will the tester have sufficient throughput capability for the volumes of product expected? Estimates of test time are difficult to produce with any accuracy, and many tester manufacturers publish limited timing information. The test engineer must carefully evaluate the throughput information, as purchasing a high speed machine may not be cost effective. For example, we will examine a hypothetical manufacturing process that expects to produce 1,000 units per month and two competing testers, one that can test a unit in 20 min and another that can test the unit in 15 min. For the purposes of this discussion, we will define the test time as encompassing the loading and unloading time and any expected retesting and maintenance. The first tester will require 20,000 min to test 1 month's worth of product, while the second machine' only takes 15,000 minutes to test all 1,000 units in a month. With a 4.5-week month, the first machine uses nearly 75 hr/week for testing, and the second machine uses almost 56 hr every week. Business issues come into play here, and may justify the purchase of more than one machine or the use of double shift manufacturing or overtime. These decisions are normally out of the hands of the test engineer. The relative costs of the machines would enter into the decision as well as a host of other considerations mentioned below.

There are more dimensions to the selection process than throughput and electrical specifications. Each machine has specific MTBF profiles, as well as a requirement for periodic calibration and maintenance. The frequency or complexity of calibration procedures may negatively influence the purchase of a tester with otherwise stellar specifications. Large manufacturers are able to fund internal calibration and maintenance department by virtue of the costs associated with downtime on a large number of test systems. Smaller manufacturers do not have the luxury of captive maintenance and calibration forces, and rely upon the equipment vendors or independent calibration and repair laboratories. Maintenance and calibration may consume valuable time that would otherwise be spent producing revenue generating product, so many manufacturing plants purchase duplicate instruments or tester subassemblies that can be pressed into service while the primary units are being repaired or adjusted. These spare subsystems are expensive to purchase and maintain, but their cost is offset by their ability to reduce the effects of tester failures upon product volume. The test engineer works with the tester manufacturer to determine the appropriate number and type of spare instruments and parts to purchase, so that manufacturing funds are put to their best use. The line between technical and business decisions is fuzzy at best, and a generous amount of cooperation must exist between the person who evaluates technical issues and the person who examines business considerations.

Software and Human Interface Concerns

In automated testers such as an ICT, overlooking the actual usability of the product to write and maintain test programs can be a very real danger. This can become a big factor in making a decision. For instance, this author believes that one of the big features of the GenRad ICTs is the user-friendliness of the software. GenRad wrote the software so that test engineers can easily write and support test programs. A competitors similar system was almost unusable to demonstrate anything other than canned test programs. Make sure that automated systems are carefully evaluated from every angle to prevent future difficulties in using the product. When possible, write an actual test program on the leading contenders and then compare not only the resulting program, but the ease of handling board level engineering change orders (ECOs) and rewriting each type of test.

Business Considerations

Money is the asset behind all business decisions. Business issues look at the return to be expected from an investment . A dollar spent should generate a dollar or more of revenue, otherwise the process operates at a net loss. The return on investment (ROI) may be rather indirect, for example, using a slightly more expensive component may double the product reliability, and conversely a part with slightly lower cost may have little or no effect upon the product reliability. Such an indirect benefit may be difficult to measure, and may be significant only in statistically large unit populations. A strategy that results in a cost that equals or exceeds the expected benefit is not good business. Often business decisions are driven by governmental regulations, such as the avoidance of inexpensive but illegal processing chemicals or techniques. The economic benefit of using illegal methods is offset by the prospect of fines or unflattering press that result from prosecution. Penalties for illegal activities are selected to produce a negative return on investment and thus influence the business decision. Parameters of test that affect business decisions are those which affect the return on investment: cost of a tester, potential capacity, useful life as compared to the depreciation schedule, amount of human support needed, maintenance cost, MTBF, cost of spare parts, cost of calibration, floor space required, plant support (power, lighting, water, cooling, vacuum, etc.) required, value of the machine on the used equipment market, cost of equipment leasing, operator training, test development costs, percent utilization, and incremental costs of upgrading. The multi-dimensional problem that confronts the person charged with making the business decision may be daunting. The final decision must be an amalgam of the technical and business decisions, as a machine chosen purely for technical reasons may make poor business sense and a tester selected for its attractive business characteristics may be of little or no use on the factory floor. The decision is facilitated where a single person has the business acumen and technical expertise to evaluate competing test alternatives. The decision takes much longer and has more complexity as more people are added to the approval process, because conflicts in perception between people often make people proceed more cautiously than they otherwise would. A test system selected by a large group of people often costs more and has specifications that far exceed those found in a machine chosen by an individual with business and technical responsibility. A test engineer without direct business influence does have the ability to influence the business decision by researching the criteria that drive the business decisions within a company. Most corporate business decisions are motivated by the composite return on investment, that is how much money will be made by making a certain purchase. The costs and monetary benefits are elusive to calculate, but possible to determine. The payback per dollar spent is the factor which will be most influential in a nontechnical setting. It is important to avoid confusing payback in revenue and payback in imponderables which do not affect the corporate balance sheet (peace of mind, technical superiority, etc.). A careful analysis of the financial issues that accompany a technical comparison will make a positive impression on the people who make the final business decision.

12

Choosing The Right Tests

Types of Testing

In order to choose the correct set of tests, one must first gain an understanding of the reasons why the testing is to be performed, and under what circumstances. Let's examine some of the reasons for testing a power product:

- **Engineering Test**—The testing of a design to specifications; or, in other words, testing a design to verify that it meets all specification.
- **Component Test**—Testing individual components as purchased or manufactured, for the purpose of either qualifying a component or verifying component quality at incoming inspection or on a board or assembly, to detect manufacturing process errors.
- **Component Burn-In**—Burning-in individual components prior to assembly operations to accelerate the occurrence of manufacturing defects and to locate them before they can be assembled into a product.
- **In-Circuit Test**—Testing board level assemblies for assembly errors or parts that fail to meet specification.
- **First Power on Test**—A test of completed assemblies with power for component and assembly errors that cannot be located until the application of power.
- **First Product Test**—A test of completed products for some assembly process errors.
- **Test After Rework**—A test to locate faults as a result of any rework or repair actions on a product; a test to locate any faults masked by previously discovered faults.

- **Diagnostic Testing**—Testing for the express purpose of locating the exact cause of the fault or error.

- **Accelerated Life Testing**—Testing under conditions designed to provide results on the estimated life of a product.

- **Ongoing Reliability Testing**—Verification of the capability of the manufacturing process to produce products which consistently meet reliability specifications.

- **Screening Field Returns**—Testing to identify those returns that meet specifications and do not need any repair; the danger here is that some failures or faults may not be readily discovered. This is a good reason for doing additional testing for No Problem Found (NPF) field return modules. A short burn-in for these kinds of NPF modules makes a lot of sense, and is less expensive than finding out that there really was a problem after the module or system is put back into service at a customer's site.

- **Test After Repair**—Testing to verify that the repair process did not insert any faults into the product, and that all faults have been repaired.

Classes of Testing

Testing falls into three distinct classes: Engineering Testing, Manufacturing Testing, and Field Testing. Each of the classes has different goals in testing power products.

Engineering Testing

Engineering Testing is testing done in, and usually by, engineering to test a design concept, circuit design, or the suitability of a component or part for a design. Engineering testing can be much more involved than manufacturing testing. During some qualification testing, parts or circuits may actually be tested to the point of failure. The goal of test here is to validate a design. It can be very cost effective to perform simulated testing, discussed in Chapter 2, to determine that during manufacturing all correctly assembled parts will work together as desired. Design verification testing can be performed using the analog workbench, which can actually accept inputs such as desired bode plots and following simulation, determine the best set of component values to produce the desired plots. This can be very cost effective. I have seen production plants where a "small" percentage of the product produced (5–15%) failed to operate due to component tolerance buildup. These conditions must be eliminated before manufacturing receives or accepts a product design and starts to ship it.

Manufacturing Testing

Manufacturing Testing is testing done during production and usually by production personnel to test for manufacturing process faults. Manufacturing testing is usually much less involved than testing done by engineering personnel. One of the goals of manufacturing testing is to never stress a part or product during testing to the point where that part or product may be damaged in any way. The goal of testing here is to validate the production processes.

The only purpose of manufacturing testing is to provide timely feedback to the manufacturing processes to correct faults in the process where product is allowed through the process with defects. If this attitude is taken from the start of a new product then by the time the volumes reach significant numbers, all of the process faults will have been removed and only good product will be manufactured. The purpose of testing is to fix the manufacturing process so that product is manufactured error free 100% of the time. If the process is designed such that it allows errors, then they soon become acceptable. They are not acceptable. In manufacturing power supplies, it is possible to achieve very high quality levels and lower costs by creating a process where any faulty unit or assembly or component, is sent to manufacturing engineering. Manufacturing engineering's job is then to determine how this fault happened, and how the process let it get through. Then they can modify the process to prevent any reoccurances of that fault. The process then becomes a living, changing, improving process. It is not fixed, like a document filed away. It changes daily.

Field Testing

Field testing is testing done after a product has failed in the field or at least is suspected as having failed or caused a failure in a system in the field. Field testing can be performed by field service, production personnel, or even engineering personnel to determine the true reason for the fault. One consideration that helps to determine the level of test is the point in the product life cycle where the fault occurs. Field testing is often somewhat less involved than testing done by manufacturing and engineering. One of the goals of field testing is to test those component parts of a power supply that deteriorate relatively quickly with age, such as electrolytic capacitors. The goal of testing here is to validate that when the product has been repaired that it will be as good as new from a operational point of view. It is for this reason that field service testing may include a few additional tests not performed elsewhere on the product when it was new.

Goal of Testing

The real goal of testing is to make testing obsolete and unnecessary. To do this a process must produce timely and useful feedback to the design and manufacturing processes for continual product and process improvement. Testing a product just to repair it for shipment is a mistake. The purpose of testing is to locate the fault in the process that allowed a product to be produced that did not meet specification.

The Menu of Tests

The following tests have been provided as a menu to choose from. Each power supply may require different tests due to circuit topology and/or design margins. Some tests will be dropped after some level of experience testing a product has been gained. Process improvement will also negate the need for some tests over time.

Screening Tests

The next group of tests are made prior to the application of primary power.

- FIXTURE CHECK To check that the model number of the power supply fixture is the same as the test program expects.
- BACKPOWER TESTS
  - OUTPUT BACKPOWER TEST—To check the outputs for shorts prior to power being applied to the supply.
  - BIAS BACKPOWER TEST—To check that a short doesn't exist on the bias supplies.
  - REFERENCE VOLTAGE CHECKS—To check chip reference voltages prior to power being applied to the supply.
  - OVERVOLTAGE TRIP POINT TEST—To check each outputs overvoltage trip point prior to power being applied.
  - OVERVOLTAGE TRIP POINT RECOVERY—To make sure the outputs recover from having an overvoltage applied to the output.

* OVERCURRENT FUNCTIONALITY CHECK—This check is to make sure the overcurrent circuitry is functioning prior to primary power being applied.

* OVER POWER FUNCTIONALITY CHECK—This check is to make sure the overpower circuitry is functioning prior to the application of primary power.

* OSCILLATOR FREQUENCY CHECK—This test checks the frequency of the Oscillator in switching regulators.

* PULSE WIDTH MODULATOR CHECK—This test checks the pulse width of pulse width modulator (PWM).

* EXTERNAL SYNC CHECK—The purpose of this test is to check that the external sync is working properly.

* EXTERNAL FREQUENCY SHIFT TEST—The purpose of this test is to check the external frequency shift is working prior to power being applied.

* SOFTSTART CHECK—The purpose of this test is to check that the softstart functionality is working prior to power being applied to the supply.

* MAIN SWITCH TRANSISTOR SHORT TEST—The purpose of this test is to check that a short doesn't exist on main switching transistors prior to power being applied to the supply.

NOTE

THE FOLLOWING SET OF FUNCTIONAL TESTS ARE MADE WITH PRIMARY POWER APPLIED. USE **EXTREME CAUTION** and **TRAINED PERSONNEL ONLY.**

---

* INPUT CURRENT MONITOR (DURING RAMPUP AT MINIMUM LOAD)—This test checks that an AC short doesn't exist in or on the supply prior to the application of primary power. This test also checks for excessive primary current that may be caused by filter capacitors which have the incorrect voltage rating, or by input rectifier problems. Some switching power supplies may require that the loads be in the constant resistance mode during this test.

* AC INPUT RMS CURRENT TEST—The purpose of this test is to check the AC RMS input current value.

* AC INPUT PEAK CURRENT TEST—Checks the PEAK AC input current at turn-on. For this test the point of turn-on must be specified as to voltage and the point in the cycle that turn-on occurs.

- **AC INPUT RECURRENT PEAK TEST**—The purpose of this test is to check the AC recurrent Peak input current.

- **HIGH LINE & MINIMUM LOAD CHECK**—The purpose of this test is to check regulation at the point of high line and minimum load. On some supplies the switch position 130–260 V will have to be checked.

- **LOW LINE & FULL LOAD CHECK**—The purpose of this test is to check regulation at the point of Low Line and Full Load. On some supplies the switch position of the 115–230 V or 130–260 V switch will have to be checked.

- **DYNAMIC LOAD TEST OR LOOP RESPONSE TEST**—The purpose of this test is to test the response of a power supply to a step response or a change in load at a specified rate. The purpose of the test is to verify the large signal AC characteristics of power supply. Typically this is done with a 10–90% load current step change at a specified rate such as 10 A/μs. Supplies designed to fill less demanding applications may be tested to a less demanding specification.

- **AC SOURCE FREQUENCY CHECK**—The purpose of this test is to check the power supply operation at the minimum and maximum frequencies of input power. Usually 47–63 Hz, although many supplies are specified to operate at 40–1000 Hz.

- **PERIODIC and RANDOM DEVIATION (PARD) TEST** is a well-defined method of performing ripple and noise checks at the outputs of a supply.

- **RIPPLE and NOISE**—There are many ways to measure ripple and noise including: with a Peak-to-Peak Detector, with an oscilloscope, or with a voltmeter. The problem with these methods is that they each have drawbacks, and it is very difficult to measure to specification, unless the specification includes the details of how the test is to be performed. This author feels that a better way would be to measure ripple and noise with an automated instrument or as part of an automated power supply test system in the following manner, divide the frequency band into two or three components, defined as follows:

  * For **Linear Power Supplies:**
    - **RIPPLE** 45 Hz to 10 KHz, periodically related to the line frequency.
    * **NOISE** 45 Hz to 30 MHz, not periodically related to the line frequency.

  * For **Switching Power Supplies and Inverters:**
    - **LINE RIPPLE** 45 Hz to 10 KHz, periodically related to the line frequency.

* **SWITCHING RIPPLE** 10 KHz to 10 times the switching rate, periodically related to the rise and fall of the switching.

* **NOISE** 45 Hz 30 MHz, not periodically related to either the line frequency or the switching rate.

As can be seen, this is not a simple instrument, but one where the source of the ripple or noise is clearly identified. This instrument does not exist today as far as I know, but it is my wish that some enterprising instrument company or power supply tester manufacturer will see the need for this type of instrument and produce it.

* **OVERCURRENT CHECK**—To check that the power supply responds within the specified overcurrent trip points. Quite often a response time is indicated as well.

* **OVERCURRENT RECOVERY**—The purpose of this test is to make sure the power supply outputs will return to regulation after an overcurrent condition.

* **SHORT CIRCUIT CHECK**—The purpose of this test is to check the short current capabilities of the power supply outputs to specifications. This test must not be made on any output that is not protected in some way.

* **SHORT CIRCUIT RECOVERY CHECK**—The purpose of this test is to check the power supplies will regulate after a short has been applied to the output. This test must not be made on any output that is not protected in some way.

* **DYNAMIC LOADING TESTS**—The purpose of this test is to check the closed loop small signal response of the regulation amplifiers in a power supply. Typically this is done with a 10% variation of the load current with a step change specified at a rate such as 10 A/;gmS. Supplies designed to fill less demanding applications may be tested to a less demanding specification.

* **TIMING TESTS**—The purpose of this test is to check the timing relationship between the various AC and DC signals that may be generated by the power supply such as: AC OK, AC Low, DC Low, DC OK, Power OK, Power Fail, etc.

* **LED OUTPUT OK CHECKS**—The purpose of this test is to check all LED indicators are functioning.

* **SUB FUNCTIONAL BLOCK TESTING**—This type of testing involves the testing of small groups of components below the functional block level in order to give further diagnostic direction as to the cause of a given malfunction.

Menu of Tests Available

Input Functional Testing

- AC INPUT TESTS
  - * HIGH LINE
  - * LOW LINE
  - * INRUSH CURRENT
  - * POWER FACTOR
  - * STATIC LINE REGULATION
  - * DYNAMIC LINE REGULATION
  - * STARTUP TIME
  - * HOLDUP TIME
  - * DROPOUT TIME
  - * LEAKAGE CURRENT
  - * MILITARY SPECIFICATION TESTING
  - * OTHER TESTS (per Spec.)
- DC INPUT TESTS
  - * HI LINE
  - * LO LINE
  - * INRUSH CURRENT
  - * SOURCE RIPPLE and NOISE
  - * STATIC LINE REGULATION
  - * DYNAMIC LINE REGULATION
  - * STARTUP TIME
  - * HOLDUP TIME
  - * DROPOUT TIME
  - * LEAKAGE CURRENT
  - * MILITARY SPECIFICATION TESTING
  - * OTHER TESTS (per Spec)
- CONTROL INPUT TESTS
  - * POWER ENABLE
  - * POWER DISABLE
  - * MARGINING
  - * MASTER CLOCK INPUT

Output Functional Testing

- DC OUTPUT TESTS
  * OUTPUT VOLTAGE
  * STATIC LOAD REGULATION
  * DYNAMIC LOAD REGULATION
  * SHORT TERM DRIFT (voltage)
  * LONG TERM DRIFT (voltage)
  * THERMAL DRIFT (voltage)
  * RIPPLE and NOISE
  * RIPPLE
  * NOISE
  * OVERVOLTAGE
  * CROSS REGULATION
  * CURRENT LIMIT
  * SHORT CIRCUIT

Isolation

- DC
- AC
- NOISE
- EMI/RFI

AC Output Tests

- STATIC LOAD REGULATION
- DYNAMIC LOAD REGULATION
- FREQUENCY STABILITY
  * Versus time
  * Versus temperature changes
  * Versus input voltage changes
  * Versus load current changes

- WAVEFORM DISTORTION
- SPECTRAL PURITY
- SHORT TERM DRIFT (voltage)
- LONG TERM DRIFT (voltage)
- THERMAL DRIFT (voltage)
- RIPPLE and NOISE
  * RIPPLE
  * NOISE
- OVER-VOLTAGE
- ISOLATION
- CROSS REGULATION
- CURRENT LIMIT
- SHORT CIRCUIT

Ancillary Testing

- AC OK
- DC OK
- Power Sequencing
- Real Time Line Clock
- MILITARY SPECIFICATIONS TESTING
- OTHER SIGNAL OUTPUTS

EMI/RFI Testing

- ON SITE
  * BENCH TESTING
  * SCREEN ROOM TESTING
  * CONDUCTED EMISSIONS
  * RADIATED EMISSIONS
  * MILITARY SPECIFICATIONS TESTING
- OFF SITE
  * CHOOSING A SITE
  * SPECIFYING TESTS
  * INTERPRETING RESULTS
  * DOCUMENTING RESULTS

- PRE-PRODUCTION TESTING
  - * PILOT PARTS SPECIFICATIONS
  - * PILOT PARTS TESTING
  - * PILOT RUN FEEDBACK
- LIFE TESTING
  - * LIFE TESTING
  - * ACCELERATED LIFE TESTING
- EQUIPMENT NEEDED
- PERFORMING THE TESTS
- INTERPRETING THE DATA

The Selection of the Appropriate Tests

Task selection is a part of the process definition and as such is not fixed in time, but varies with the needs of the process. At new product startup, it is normal to have a large number of tests as the process is gradually debugged with increasing volumes and quality. As the need for a test is diminished, it will reach a point where the appropriate action will be to remove that test from the list of tests. By the time that full production is in motion, the list of tests will have been greatly reduced. Throughout the life of the process, for a variety of reasons, tests will be added and then deleted from the process. This is normal for a live process.

The Appropriate Ordering of Tests

Tests tend to be grouped into the following categories for board level in-circuit and quasi-functional module testing:

- Continuity Tests
- Passive Component Tests
- Active Component Tests
- Custom Component Tests
- Untestable Component Testing
- Loop Testing

Tests tend to be grouped into the following categories for functional power supply testing:

- Prepower Testing (Safety Tests)
- Initial Power Testing
- Light Load Testing
- Full Load Testing
- Power Cycling.
- Discharge (Safety Testing)

Inside the groupings, test ordering is most often not of much concern, except for those cases where one test sets up the following test. This can save much time by reducing redundant setups.

13

Component Testing

The testing that occurs earliest in a manufacturing process is related to component testing. One manufacturing philosophy asserts that if one perfectly assembles perfect parts, perfect products will result. Quality control has two missions: to ensure that parts are as perfect as possible, and that the process of assembly introduces no errors.

Component testing fulfills the first mission. By its definition, assembly combines several individual parts into a larger unit. Component testing must examine the large quantities and broad variety of parts prior to assembly. Testing parts will consume considerable time and money because the handling and test expense increase the final cost of every part. An assembly may contain dozens or hundreds of parts, and the trivial cost per part becomes a very significant portion of the product cost when multiplied by that volume. One factor that aggravates the cost effects of component testing is the decreasing cost of components.

Electronic components have decreased in price steadily since the early 1960s. The cost of testing has not dropped as quickly, so that the final price of a part may be based more on the cost of test than on the cost of raw materials. Competitive pressure has placed emphasis on making test more efficient and less costly.

Ship to Stock

Ship to stock is the name given to a technique for controlling component testing costs. This method makes the component supplier responsible for the quality of the components used by the customer.

The customer negotiates a contract that specifies the important parameters of the components, the methods the supplier will use to guarantee that the components meet those specifications, and the penalties for nonconformance. The supplier will charge slightly higher than normal prices for the components to compensate for the additional testing, handling, and care that will be exercised during manufacture of the parts. The customer pays a small premium on each part for the ability to transfer the responsibility for component quality testing.

An interesting anomaly occurs when making statistical analysis of components of medium tolerance. Most manufacturers processes have a Gaussian distribution around a tolerance mean (See Figure 13-1). The manufacturer may measure the parameter of interest and divert components within a tight tolerance range to a finicky customer, and deliver the residue to other customers with lesser requirements. The distribution of tolerance seen by the second tier of customers has a notch in the center (or some other part) of the tolerance range (Figure 13-2). This deviation from the expected distribution curve is additional reason to perform worst case design, as the potential exists for every component to have a value near the limits of its tolerance. The moral of the story is to ask for what you need, and not to assume that you will get the average product. You get what you pay for.

The supplier normally performs some testing on the components for every customer, so the ship to stock concept involves little more than a few additional tests, improved record keeping, and a bit more quality control.

The customer may require the use of specific processes and tools for monitoring quality, and a level of quality may be necessary that exceeds the

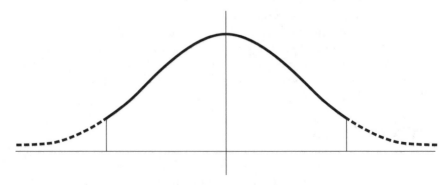

Figure 13-1. Gaussian Distribution and Mean

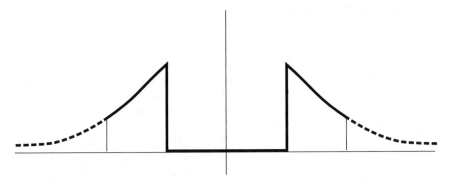

Figure 13-2. Gaussian Distribution after Part Skimming

normal standards of the supplier. The customer and supplier must cooperate to achieve the mutual benefits of a ship to stock arrangement, as an adversarial relationship results in higher costs, delays, and mistrust. The customer may not trust the vendor to conscientiously follow the proscribed test regimen. The supplier may not trust the customer to purchase enough parts to offset the considerable cost of negotiations, planning, and setup.

A ship to stock agreement may contain provisions for reviewing the performance of the vendor. The customer reports the actual quality level of the components as received. The customer may test random lots of the components from the supplier and measure quality from the results of that test, or may record the nature of each fault that is traced to a faulty component from the vendor.

A supplier may be evaluated, and the results of the manufacturing process closely examined to determine if the methods of manufacture and quality control are sufficient to assure that a specific level of good components will be produced. This is termed process qualification. Considerable detail of the vendor manufacturing and quality control processes may be required, and this demands cooperation from the vendor. If the customer has specific needs that are not covered by the vendor in a standard product, process qualification allows the customer to predict the performance of the vendor's process. Process qualification facilitates the purchase of new components similar to other parts made in the qualified process.

A quantity of a component from a vendor may be purchased and tested to measure the level of quality. A product with sufficiently high quality is considered a qualified product, and may be purchased from the vendor without further investigation. This analysis not require that the details of the vendor process be known, but is specific to a particular product and not to a product family or process. Product qualification is generally simple, quick and low in cost compared to process qualification.

Third Party Test

Another method of component testing is to contract a third-party to test components. The independent company may spread the testing overhead among several clients and control costs. This option is particularly attractive when the supplier and the customer have insufficient to perform the tests in an economic manner. In high volume the handling and overhead of the third-party may cancel money savings.

Third-party testing is advantageous as an emergency strategy when test equipment at the vendor or customer is disabled or damaged and product is otherwise ready for delivery. The occasion of such an emergency is not the time to begin planning for the event, so alternative test strategies should be examined and plans in place for most critical components.

The processes used by a vendor may be examined in a direct or indirect manner. Direct monitoring involves periodic inspections of the vendor process to confirm that no changes have been made to the process, removing or adding process steps, or otherwise decreasing the percentage of acceptable components delivered from the process. The customer must stipulate the conditions surrounding the inspections: the frequency, type, and amount of notification prior to an unscheduled visit, the process and product parameters that will undergo scrutiny during the inspection. These conditions must be spelled out in the purchase contract for the components as a condition of the sale. The burden of proof of nonperformance in these cases usually falls to the customer, and the vendor is under no responsibility to aid in the collection of evidence. The penalties under these contracts are virtually all directed toward the supplier for nonperformance, should the customer determine that the contract has been breached. These agreements are customarily steeped in legal language that is understood best by lawyers.

Indirect monitoring of a vendor process is accomplished by frequent testing of random samples of the product delivered to the customer. The vendor is not a party to the analysis as in direct monitoring, and the customer is responsible for the entire analysis. The customer must report the details of their investigation to the supplier when seeking compensation for noncompliance to the purchase contract. The supplier may elect to verify the claims of the customer by retesting the components found faulty. In general, the customer receives less compensation for faulty components found by indirect monitoring than the same number of faulty parts uncovered in direct monitoring scenarios. The benefits of indirect monitoring are that no agreement is needed between the vendor and client about inspection, and the analysis of faulty components may be a logical extension of the customer testing processes in contrast to the time and travel required to perform a supplier plant inspection.

Incoming Inspection

Components purchased in less formal situations without purchase agreements obviate the need for process or product qualification. Parts purchased from distributors may include parts from a variety of vendors merged into a single shipment. The customer must test these parts when they are delivered. This testing is termed incoming inspection.

Parts may be tested on a random sample basis and statistical methods applied to determine the acceptability or nonacceptability of the lot. In other more stringent situations, every part received is subjected to the test suite. The purpose of incoming inspection defines whether sampling is performed or all parts are tested.

If the intent is to evaluate the risk of using a particular shipment, then sampling is appropriate. Using ship-to-stock agreements it is possible to reject an entire lot of components based upon the statistical analysis of failures detected in a random sample from the lot. The tests performed on randomly sampled components are exhaustive: measuring every parameter in the specification of the component because virtually any deviation from the accepted list of specifications will render a component unacceptable.

A test of the key specification or a small subset of the specifications of a component is done when every part in a lot is tested. The test may concentrate on the parameters that are of greatest interest such as the capacitance of a capacitor in deference to its leakage or equivalent series inductance (ESL). In some cases, the incoming inspection may place emphasis on parameters that have been marginal or out of tolerance in earlier shipments, indicating instability in the manufacturers process.

Components that pass incoming inspection are not returned to the supplier or reported as failures to the supplier if they fail later in the manufacturing process. Completion of incoming inspection constitutes acceptance of the shipment in most purchase agreements, so the onus of sufficient incoming inspection is the burden of the customer.

The testers used in incoming inspection stations are designed for two classes of components: sorted and unsorted. Sorted components may be on tape reels or in shipping tubes which hold the components in very predictable positions. Unsorted components are shipped in bags or boxes and must be physically separated and positioned before testing.

Component testers are designed to position each component within the test environment and test it as rapidly, completely and accurately as possible. Some testers are designed for resistors, inductors, and capacitors, while others are designed for the specific needs of semiconductor testing. The complexity of semiconductor testers limits their throughput and increases the amount of time required to perform the setup. Semiconductor testers measure a broad variety of devices and parameters of those devices, ranging from simple diodes to large-scale ICs. The cost of testers with broad capabilities is relatively high, but this cost pales in comparison to the price of a series of specialized testers.

The responsibility of the test engineer is to carefully define the incoming inspection tests in terms of parameters of interest, records to be maintained, and protocol when faults are detected in a particular component or class of component. The selection of testers is often dictated by the parameters being tested, the volume of components to be processed, and the applicability of the tester to the manufacturing process. A primary consideration is the ownership of existing equipment, as incoming inspection component testing must be cost effective. Existing equipment in a manufacturing line is very economical when compared to the cost of new capital equipment. The use of existing equipment in cost-sensitive production lines may affect the choice of components selected for a particular design, and rarely will a single new component or group of components mandate the purchase of a new tester for incoming inspection. A new tester for incoming inspection may be desirable if the volume of component would overwhelm existing systems, or if critical parameters are untestable with the current setup.

The feedback from the incoming inspection process goes to the purchasing department as leverage for future purchases, and to the supplier (if required by the purchase agreement) as a means of process control. Purchasing engineers will be interested in the numbers of unacceptable components, while the supplier may wish to know the actual degree of error detected. On larger and more costly components such as ICs and subassemblies, the serial number of a faulty component may contain additional information not contained in the lot number or ship date.

Test During Insertion

Some machines used to insert components onto modules have the option of testing the parts just before they are placed on the modules. This test is no substitute for incoming inspection, but it can be used to verify that the proper parts have been loaded into the machine before a soldering operation makes removal and replacement of erroneous components an expensive and time-consuming process. Because this test is not expected to establish the quality of a component for acceptance reasons, and is specifically designed to prevent improper assembly, the goal is to detect a component with the wrong primary parameter. Many of the insertion testers cannot distinguish the difference between similar semiconductors, and are unable to distinguish components that have similar primary values, such as resistors of the same value but differing power rating or capacitors of the same value but differing voltage rating.

Some insertion machines use tapes that contain a variety of components in the proper sequence to be placed on the module. The default action of these testers is to reject the component and to identify the assembly as improperly assembled because of a missing component. Insertion machines that contain a variety of identical components on individual reels or input bins will reject those components that do not meet the insertion criteria, and continue pulling parts from the reel or bin until an acceptable component is found. Intelligent machines abandon the search for acceptable parts if a preset number of bad parts are detected in sequence with the assumption that the machine was loaded incorrectly. In these cases, the human operator of the insertion machine must remove the incorrect parts from the machine and then load the proper reel or bin of parts.

The tests performed during insertion are necessarily simple due to the limited time for making the test as the part is positioned for insertion into the module. Resistors are measured for their ohmic value. Capacitors are measured to determine their capacitance value. Transformers and inductors are tested for continuity through the windings and (where time and the capability of the tester allows) the inductance of each winding is measured.

Testing active components during insertion presents special problems. The differences between one transistor and another or one IC and another may be subtle. Diodes are usually tested for their ability to conduct in one direction and not in the other during insertion. Additional tests such as reverse breakdown and recovery time usually require sophistication that is not available in the test equipment found at insertion stations. Transistors are tested for polarity and identification of leads. The insertion test generally does not include gain or breakdown tests which might differentiate one transistor from another. The insertion test will detect transistors which are physically different, of the wrong polarity [NPN versus PNP], or of the wrong type, [eg., junction field effect transistor (JFET) as opposed to bipolar or to metal oxide semiconductor (MOS)].

Integrated circuits can be more difficult to test during insertion. The test equipment used for testing multiple pin devices is physically complex, electrically intricate, and expensive. Testing ICs during insertion becomes cost effective when the frequency of errors in part orientation or part selection raises the cost of correcting those errors above the cost of test capability.

Some component types and certain values of components require such specialized equipment that testing during insertion is virtually impossible. Ferrite beads are difficult to test, as their presence around a wire makes the wire act as a small valued inductor usually with a very low Q. [The Q of a coil is the ratio of inductive reactance to dc resistance.] Transformers with high turns ratios present problems because of the high voltages which are produced. Thermistors can be tested for some degree of conductivity, but temperature control during insertion is usually poor, so characterization of a specific thermistor value or transfer characteristic is difficult to perform. One obvious component that defies testing is a fuse: it will only operate correctly one time, so testing for value is impossible.

Insertion machines may include test equipment within the body of the machines themselves. Testers found in commercially available insertion machines are able to measure resistance, capacitance, inductance and continuity. Testers with the ability to measure semiconductors and ICs are found in custom designed insertion systems. The market for complex testers is very small, and therefore this capability is not often found in commercially available equipment.

Complex testers contain a configurable physical interface, so that a broad variety of components with differing physical attributes can be tested. A switching system may be employed so that additional test instruments may be connected to the physical interface. The, physical interface is a pressure contact or zero insertion force interconnect, so that the insertion mechanism positions the component in question against the interface, the test is performed, and the component is inserted in the module (if it passes the test). The insertion motion system, test instruments and interface signal switcher are invariably controlled by a programmable controller, processor, or computer. The primary criteria for the test equipment are low cost and speed, while accuracy and precision are relatively unimportant.

The choice of equipment may depend upon other factors such as the types of parameters and range of measurements, the ease of programming, or the number of signal pins that may be controlled. Much of the cost of the test equipment is involved in the signal switching and physical interface.

Reliability of contact and the interconnect quality can have a profound effect on the measurements performed by the test equipment. It is most important to remember that testing during insertion is not a test of the component quality as much as it is a test of the presence, orientation, and identity of a component. The insertion tester does not test the goodness of the parts, but that they are the proper parts inserted in the proper places in the correct orientation.

14

Module Level Testing

One image that comes to mind when testing electronic circuits in a factory environment is the picture of module testing. In reality, there are several types of module tests that may be performed in addition to the familiar pass-fail testing of each module. We will examine in-circuit and functional testing in this chapter.

In-Circuit Testing

Power supply modules contain circuits with considerably larger energy content and destructive power than is found in logic circuits or many other analog modules. A test that determines if gross wiring or component errors exist before applying normal power is used to prevent premature destruction of improperly assembled modules. In-circuit testing satisfies the goals of such a low energy test.

In-circuit testing uses a specialized tester to electrically test each node of the module for short circuits to adjacent nodes, and to determine if the correct components are in place and firmly connected. The soldering operation between the component insertion step and the in-circuit test may damage or dislodge components, or the soldering operation may be incomplete in a region of the module causing opens, or introduce short circuits via solder bridges in another. The in-circuit testing process has the goal to detect assembly errors and provide feedback to the process to correct the cause of those errors. Typical errors detected during in-circuit test may include:

- Missing and wrong components
- Shorted etch and components
- Poorly soldered or open component leads
- Components of improper value not discovered earlier
- Components that were damaged during insertion or soldering
- Components with improper orientation

In-circuit testing is a most important process gate due to the fact that most assembly errors can be detected at this process step with the correct tester. With the inclusion of loop testing as a part of the in-circuit test process, modules that pass the this process step are virtually certain to operate when assembled into a finished unit. When used in this manner, direct labor costs of diagnosis can be severely reduced. This importance allows the allocation of a larger percentage of the testing budget to this process step. One important aspect of the tester specification is it's ability to test a broad variety of components in the range of each of them. Power supply modules usually contain resistors in values less than an ohm while other resistors may have values in excess of a megohm. Capacitors found on power supply modules may range from a few picofarads to a nearly a farad. Transistors may be JFETs, MOSFETs, or bipolar and may be rated from a few milliwatts dissipation to several hundred watts. Integrated circuits are likely to be a combination of operational amplifiers, comparators, references, and power amplifiers as well as some logic circuitry in TTL or CMOS employing such functions as flip-flops, AND gates, OR gates, XOR gates, Schmidt triggers, inverters, and buffers. Transformers and inductors may have windings of one turn to several hundred turns, and inductances ranging from nanohenries to several henries. Diodes may be used for switching, voltage regulation, transient protection, or power rectification. Special parts may include thermistors, SCRs, circuit breakers, or fuses, each with its own unique testing difficulties.

3, 4, and 6 Wire In-Circuit Testers

The tester chosen for the in-circuit test must be able to handle all or virtually all of the components that will be encountered on the modules. While there are 3-wire, 4-wire, and 6-wire in-circuit test systems avail able, careful examination of the techniques used by each has led this author to conclude that the 4-wire system such as found in the GenRad family of testers is, for power supply systems, the most logical choice from both the technical and business standpoints. My conclusion from testing and observing testing by others using 3, 4 and 6 wire systems is that the 4 wire system had the least errors, lowest measurement noise, and the greatest capability. It was and is the most flexible system to use and program. This testing also revealed that 3-wire systems are not usually adequate for testing any type of power supply modules.

The penalty for incomplete test coverage is: the potential for faults to remain in a module until later process steps where the cost of detecting and correcting those faults is much higher. The equipment for performing those tests is more complex, labor intensive, and expensive.

Loop Testing

A good in-circuit tester (ICT) will have the ability to perform limited testing of the main control loop, by applying power to the control logic and bias points as well as supplying an unregulated voltage of limited current to the regulator input. This loop testing uses limited energy in comparison to a full power test, and has little potential to damage other components while providing incontrovertible evidence that the regulator circuitry is operating to some extent. The capability for loop testing will not add significant cost to an in-circuit tester, provided that a good one was chosen originally. The addition of loop testing increases board coverage essentially to 100%. This has the impact of allowing the tester to be used for all diagnostic testing as well with the added benefit of reducing the cost of direct labor used in diagnosing bad modules. This will reduce the cost of manufacturing the product.

Test programming for in-circuit testers is usually highly automated via an Automated Test Program Generation (ATPG) software program. The programmer provides a representation of the module schematic in nodelist or graphic form, and a test generation tool evaluates the circuit topology to produce the test sequence. The output from the automatic test generator may include module adapter information that tells an electrical designer which points in the circuit should connect to which signal ports on the tester. More sophisticated tools take into account the physical design of a module to determine where likely shorts and opens between pads will occur. Some manual intervention is usually required to add or modify tests where the automated test generator becomes confused or does not have the appropriate component models in its library.

In the absence of automatic test generation tools, the programmer must determine the exact sequence of tests and which nodes in the circuit must be stimulated and which measured for each test. The specific conditions of stimulus and sensing are determined by the programmer, as well as the design of the adapter wiring. The programmer usually works closely with the circuit designer so that potential misapplications of in-circuit test do not damage the module that is being tested.

Writing loop tests is currently a manual process, because of the necessity of evaluating the circuit operation before the tests are written. The test designer or programmer must understand where the tester may safely apply power and what levels of power are appropriate for the circuit. The test designer may require a few components be added during the design of a module that will greatly simplify the test process, or allow variations in a node that was connected to a low impedance source such as the reference or supply signals. Loop tests are written to cover each component that has been declared untestable by the ATPG program, as well as the overall quasi-functional loop tests.

Each component to be tested is examined in the context of the circuit where it is found. In Figure 14-1 we see a tank circuit where the value of the resistor RI and capacitor C1 are not measurable by normal means due to the low value of the DC resistance of the coil T1 winding. But we can see that this is a tank circuit. So if we impress a constant voltage sine wave of varying frequency across it through a fixed resistance, at Point A, we can measure the center frequency of the tank, at the peak measurement at Point B.

We do this by injecting the input signal through a 50 or 100Ω resistor, located in the fixture, at point A, which is in the primary circuit. Our

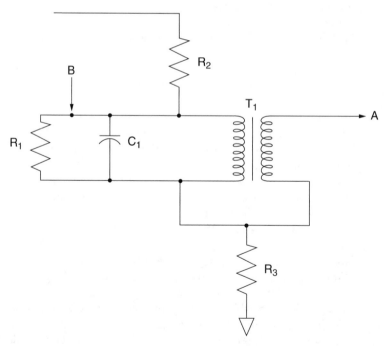

Figure 14-1. Loop Testing Example 1.

measurement node is taken at point B if it is not loaded down by the rest of the circuit, and to isolate the measurement parasitics from the tank. Further measurements made either side of the resonant frequency will yield the value of Q. Knowing the center frequency verifies the value of the capacitor, C1. Knowing the value of the Q of the circuit can verify the value of the resistor, R1. In this example the value of the resistor R1 could only be verified to within about 50%, however, that is sufficient to catch missing parts and grossly incorrect values, which are typical of the kind of process errors that are typically seen.

In Figure 14-2 we see the second example, a power transistor that may seem to be unmeasurable at first glance. The following steps provide the answer:

- Injecting a positive voltage at point A in the constant current mode set at 15 V and 50 mA.
- Grounding point D.
- Inject a square wave of at least 5 V at point C from the pulse generator. Since the generator output is impedance protected, loading it into such a low impedance will not cause a problem. Use 1 MHz to 5 MHz.
- Measure the AC value at point A. A voltage with the peak value of one half the 15 V supply should be noted.

From this example we can see that if the transistor has gain and switches, it is not installed backwards, or is not the wrong polarity part installed by error.

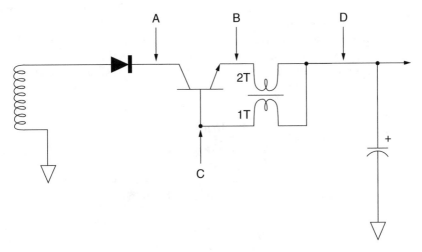

Figure 14-2. Loop Testing Example.

Next, in Figure 14-3 we see that the two capacitors appear in parallel due to the very small value of the reactance of the 10 µH choke. Using loop testing we can examine the voltage values on C7, C8, and the difference between them. Tests can be made using the internal signal generator or made during the limited operation of loop testing. Remember, we are testing to make sure that the correct component was installed correctly.

In all these examples, we have examined the component in question in light of the context of the circuit that it is located in, from a very basic point of view. Simplicity is the key here.

The design of a loop test requires experimentation with a prototype or a simulation tool (see Chapter 2). Manual generation of loop tests may take some time, until proficiency at writing them is gained and/or boilerplate models are developed. A loop test is written by applying all bias levels required for operation as well as the main input power, in DC. The input power is ramped up slowly to the point where an increase in input voltage does not cause an increase in output, indicating no-load regulation of the unit. For most power supplies, this point can be achieved with a DC input voltage of less than 100 V.

The maintenance of in-circuit test programs is minimal as long as no changes are made to the design of a circuit or module. They consist of changes to the program to add or modify a test that was missed. As production volumes increase these missed test will reduce to the near zero point. Physical changes that do not have corresponding changes to the schematic require that the test adapter be redesigned, so it is important to capture the design information for test adapters during the test design phase so that physical changes can be accommodated.

If a schematic is changed, there may be a change in the physical layout of the module. This could require a new test adapter if nodes of interest to the ICT have changed position. The test software will change as well. The circuit designer and test designer may cooperate in making changes to the test system, but usually the test engineer will be responsible for generating

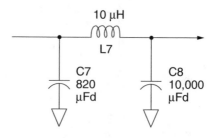

Figure 14-3.   Another Example.

the new test. The programs will have to be generated using the appropriate automated generation tool or a manual process. As mentioned above, some manual intervention may be required to make corrections to the automatically generated test, and the person performing the intervention should review the tests used before the change to determine where the original test designer had to intervene in that process.

A so-called Known Good Module (KGM) or even "golden" module, determined to operate without error, is a common method for testing new incarnations of the test program or adapter. The tester is applied to the KGM module and where the test does not agree that the module is acceptable, the test program or adapter is analyzed and corrected. Conscientious test engineers will also inject faults into other modules by removing, shorting, or modifying components to determine if the tester can detect the errors. The KGM may be applied periodically even when a change has not been made to the adapter or software just to verify that the tester continue to operate as expected, in effect "testing the tester."

In-circuit test produces a report which indicates the date and time of a test as well as the success or failure of the module. Diagnostic information indicating the identity of components that have failed or pads that are not connected to the appropriate components is also indicated. Additional information may be included on the test report such as the version of the software used and the operator identifier. The information may be sent to a video screen at the tester, and in some cases it is printed at a printer on the tester so that the "pedigree" may accompany the module as it continues in the process. Some testers have the ability to electronically transmit a copy of the report over an electronic network to an archival or analysis station away from the factory floor.

The content of the in-circuit test reports may be analyzed to determine if there is a problem with the parts being received or with the insertion or soldering steps. The in-circuit test may uncover a weakness in the incoming inspection step, or the handling between insertion and soldering. The analysis of the results is conducted by a manufacturing engineer familiar with the nuances of the process as a whole. As stated above, the in-circuit test is one of the most important tests in light of its ability to detect a large number of failure types and its chronological position in the total unit assembly process, therefore considerable emphasis is often placed on the timely and complete analysis of error trends that emerge from in-circuit and loop testing. Analysis of the results of testing is usually performed by both the manufacturing engineer and diagnostic technicians.

Some in-circuit testers use an operator-guided probe that attempts to detect the small differences in potential over the small resistances which typify short circuits and normal circuit traces. These sensitive circuits have the ability, when applied correctly, to better isolate the site of a short circuit.

Open circuits on a circuit trace are a simple matter to detect using the operator guided probe. The tester connects all but one signal path to a low-impedance reference, and connects the node of interest to a differing signal of low impedance. The operator is instructed by the control processor to place the probe over each terminus of the signal network and the test report indicates which components are properly connected to the circuit and which components are improperly disconnected. Armed with this physical information, the operator searches the etch between the connected and disconnected components until the open circuit is visually or electrically located. Opens that occur in via's conducting signals from one layer of a module to another may be difficult to find by other than electrical means, as the quality of a via is rarely ever visible.

Surface mounted components have added several complications to the in-circuit test arena. Older designs had components on one side of the module, and component leads penetrated the module and connected to the signal traces on the other side of the module. The components were considered to be on the front face or side one, and the signals were connected on the back face or side two. Thus all signals were available on the back face of the module and test adapters were composed of compression or vacuum fixtures which used spring loaded test probes to contact all of the signals on the back face of the module. Today many modules have surface-mounted components that do not penetrate the module, and may have parts mounted on both sides, complicating the connection and physical mounting of the module to avoid component damage. Several new adapter schemes have been proposed, and virtually all of them remove operator access by probing both sides of a module with large fixtures. These test adapters are electrically more complex because the operator cannot be used to compensate for the lack of a test point at one or more points in the circuit.

One consideration to examine before designing a module with parts on both sides is the impact upon in-circuit testing time, complexity, expense, and reduced diagnostic coverage. Parts on both sides may allow a more densely packed product and lower parts and assembly cost, but may significantly increase the cost of testing and fixture design.

Module Functional Testing

Module testers are designed to place a module in an environment with controlled stimulus and load, so as to measure its response to changes in those conditions. The tester attempts to simulate the conditions the module will encounter as part of a larger system while allowing the collection of measurements from the module. Most module functional testers do not include enough signal paths between the instruments and the device under test to allow in-circuit style testing, so diagnosis using module functional testers is conducted in a different manner, with much manual operator intervention. Module functional testing is used where the ICT cannot be used for loop testing. This process step usually occurs between the in-circuit tester and full functional testing. It is this process step that Loop Testing makes obsolete.

When used, the module functional tester is the final process step before the modules are assembled into larger power supply systems. The modules that fail this tester may be diagnosed at the tester in low volume operations, or at an in-circuit tester where such testing is available. Modules that pass this step successfully may be assembled into complete systems. The tester can also serve to detect any design flaws where acceptable components assembled properly on an acceptable board do not operate together to specification.

Module functional test is where many weak parts are discovered. Parts that have similar characteristics at low power as found in in-circuit and loop testing may act very differently at full power, so if an incorrect part has been used, and not detected at in-circuit testing, it will have its first opportunity to see full power at the module functional test. Some companies avoid this problem by specifying that certain components be burned in or environmentally stressed to weed out failures.

Manual versus Automated

Module functional testers are rarely as simple as a variable autotransformer, load resistor, and a voltmeter or oscilloscope. More complex modules may demand a myriad of stimulus signals and loads and therefore the tester will be correspondingly more complex. High-volume testing of complex modules will demand an automated solution. Automated module testing is performed in less time than humanly possible, so some thermal defects may not be detected at automated testers, but the ability of these systems to measure several points per second gives them the ability to detect subtle electrical defects that might escape the detection of a human at a manual functional test. The inability of automated testers to detect thermal problems may be mitigated by increased attention to the susceptible components during inspection.

Make versus Buy

Simple module designs with few inputs and outputs may be functionally tested with modest testers consisting of a few inexpensive elements. As modules become more complex, the decision of whether to buy a commercial tester or build a tester becomes problematic. The number of measurements being performed and the module complexity are the two criteria stated most often for a shift of interest from an in-house designed tester to a commercial design. Commercial testers must satisfy a broad range of customer needs to be successful, so one often purchases a tester with considerably more capability than is required to perform the job at hand. The cost of these unnecessary features must be considered when planning to purchase a tester. The equipment cost of custom built testers may be lower, but the true cost of ownership may be much higher than a commercial tester.

Tester manufacturers specialize in designing and manufacturing testers, while others must divert resources normally directed to producing products to producing testers for the product. When a custom tester fails, diagnosis and repair of the tester is borne entirely by the owner. This maintenance cost can match or exceed the savings in equipment cost over the lifetime of the tester. Tester modification and design changes can drive the cost of a self built tester beyond all expectations. A good rule of thumb to follow is that in-house tester designs are best where the tester is composed of less than $10,000 in equipment and design labor combined. This limit eliminates all automated testers except for the most primitive units.

Choosing Tester(s)

Choosing a module functional tester is a matter of comparing the capabilities of testers against the test requirements. The test requirements should include the conditions of the test including applied stimulus and power, loads, and measurements to be made. The persons selecting the tester examine requirements and machine capabilities to determine if the particular tester has the ability to accomplish the task. More than one tester may fulfill the test requirements, so other considerations become more important, such as ease of maintenance, throughput, flexibility to handle future designs, and (where automated) ease of programming. A price difference of several thousand dollars may direct the decision toward the less expensive machine, but features that make the system faster or easier to use may repay the price difference many times. The cost of spare parts or repairs can negate any purchase cost savings.

More than one tester may be purchased to increase throughput. This spreads the cost of spare parts and support of the tester hardware and software over a larger base. Automated testers require considerable debug time when a new product is introduced to them, and many companies buy a second automated tester to handle new product process debug and as a backup in case the primary tester fails. This strategy is appropriate where production occurs on several shifts and where inventory stock float is too small to absorb tester outages of more than a few days.

Factories making the transition from manual testers to automated testers should examine their product mix before writing tester requirements. A module functional tester should have the capability to handle module designs for a lifetime of 5 years. The tester can be expected to be the primary module functional tester for 5 years, so products that will be manufactured in the 5 year period after the tester is purchased will be tested on that machine.

It is just as important to avoid purchasing a tester with considerable capability beyond the current needs if the expected product mix for 5 years will not require it. In essence, do not purchase too much tester or too little.

Tests

Module functional tests are as varied as the modules themselves. We will examine the scenario of a simple one output power supply regulator module. This module might be the entire circuit of a single output open-frame power supply. We will assume that the module is a simple switching power supply with overvoltage protection on the output and foldback current limiting.

The tester would first apply a slowly increasing DC voltage of limited current into the output of the module to determine when the overvoltage protection circuit triggers. The tester would capture the peak DC value as the trigger point. The trigger point should be within the design specification, otherwise the test does not continue. The DC voltage is removed, and a pause occurs to let the over voltage circuit reset. To speed the discharge of any output capacitors, a load is applied to the output.

Next the tester would apply a nominal load to the output, and then ramp up the input power to 117 V RMS 60 Hz AC. During the first few volts of power, the current will be measured to abort the test if an over-current situation is seen to exist. After several hundred milliseconds, the operating current and power factor are measured. The power is then cycled off and then on to measure the inrush current peak value.

Designs that incorporate variable potentiometers to adjust the output voltage will be adjusted here to produce an output that is within specification. The range of control afforded by the potentiometer may be measured here if it is a part of the module specification or if the range of control will indicate a problem with the module.

The output will be monitored as the load is increased and decreased within normal limits, and the input voltage varied from low-line to high-line conditions. The input current and power factor may be monitored during these changes as well. Since the design in our example is a switching power supply, the tester would measure the output ripple and noise during the load and input excursions.

The load would be increased to determine where the overcurrent circuit triggers, then load reduced to normal levels to observe the recovery action of the module. As before, the input current and power factor may be measured during these changes to the loading.

Additional tests of the module including loop frequency and load step response time tests may be performed. These tests may uncover parasitic capacitances or inductances that would indicate weak components. Our example had a single output, but multiple output designs would employ a variety of additional tests to determine how much interaction there is between the various outputs of the module as well as the combinatorial effects of load and line disturbances. Very robust module testers may simulate the effects of missing cycles of input power, high noise levels on the power lines, or reactive load effects, just to name a few.

Marriages

Some power supply designs may require that modules be tested in sets and then shipped as a part of matched set. While usually an undesirable practice, certain design constraints may require this procedure on occasion. The modules may be physically mated before testing, or the tester may use a special interconnect for testing and the modules then physically mated after testing. Module sets are the norm where components of a circuit are spread across more than one module. Obviously, the determination of which modules make appropriate sets is impossible at the in-circuit test step without the use of loop testing. This matchmaking operation is often performed as a iterative process until an appropriate match is found.

Diagnosis

Diagnosis of modules may be performed on the functional tester in low volume applications where no in-circuit tester is available. Diagnosis on manual testers is performed by the operator, but on automated testers diagnostic analysis is performed by the control processor and may not require the participation of a skilled technician at all.

Computerized diagnosis takes many forms: the program may use functional decomposition to divide the module into small blocks of circuitry, then trace the source of errors to a particular block. Symptom directed diagnosis is a process where the tester examines the measurements taken, and by analysis of the failure signature, determines the cause of the failure. An operator may use other analytical methods, including intuition, to supplement the diagnosis of the control processor. The operator brings the ability to see, hear, and smell to the tester and may be a valuable part of the diagnostic team. A component may begin to produce a distinctive odor before any electrical degradation can be detected, or a part may glow or emit smoke without failing a test. There are cases where a switching regulator produces an unusual sound in the hours or minutes before it ultimately fails.

Diagnosis is a time-consuming and costly process, so performing this operation as early in the manufacturing operation as possible will reduce its impact. The use of loop testing on an in-circuit tester can greatly reduce the cost of diagnosing failing modules.

Field Returns

Modules that are returned from customers with failures may be diagnosed at the module functional tester, or on the in-circuit tester using loop testing. The functional tester will determine which module in a unit is faulty, and the failed module may be further diagnosed at the module tester or at the in-circuit test step. There is an important difference between new production and field returns in terms of diagnosis: new production may have missing or incorrect parts, but all of the parts should be operational. Field returns have all proper parts installed in the proper places on the module, but one or more of these parts have aged or failed. In power supplies, it is important to check the value of electrolytic capacitors on returned units for signs of aging. A repaired power supply that has a weak electrolytic capacitor is like a time bomb waiting to fail in the customer's system yet again. This difference should dictate a shift in the diagnosis strategy, although some "experts" may disagree.

15

Unit Level Testing

In examining the subject of unit functional testing we will look at the whole scope of the subject. We will define the process, examine the choice of manual or automated testing, look at the process of choosing a tester, show how to reach a make or buy decision for the actual tester and finally examine the tests themselves. For single board power supplies it may be possible to perform all the testing in one pass, where the tester has both in-circuit test (ICT) capability and functional power supply test capability. One company in fact designed and sold such a unit but for a variety of reasons did not remain in the power supply test market for long.

Uses of Functional Testing

Some ask the question, If all good correct parts are assembled correctly into a product, why is there a need to test it? The answer is that the process used to manufacture each of the components and the power supply itself are not perfect. At various points along each of the processes, small excursions from the original design specification, assuming that it was a perfect design, creep into components and products. Some are minor tolerance variations which may, as processes and manufacturing systems age, build up to make a potentially faulty part. The part may work in some applications while not in others, or it may just randomly fail in any application. In any part or assembly of parts, where human beings are involved, there will always be some imperfection present. It is the nature of human beings to be imperfect while striving, more or less, for perfection. Therefore, while we strive for perfection in our products manufactured in an imperfect world by imperfect beings with imperfect processes, we must test for the perfection we seek. In those instances where we fail to make a perfect product, we must test and diagnose to find the point in the process where the root failure occurred, and then fix the process so that the level of perfection in our products will increase. Only by striving to continually increase the level of perfection in our processes will the level of perfection in our products increase. So will our profits increase as we manufacture only good product. Should we ever attain a perfect process, testing then becomes unnecessary, to some degree and only need be used to check that the process is still perfect.

Defining the Process

There is no one perfect process that will suit the needs of all design and manufacturing processes. Each process must be designed for the products, volumes, technologies, and topologies used in the design of the product. There are two key characteristics of every process that will unlock the possibility for the success of that process becoming as perfect as it can be. The first key is the concept of a live process. By "live" we mean that it is not fixed in concrete, but like a living organism can grow, change, and adapt as the process matures. The second key is the concept of using continuous feedback to modify and improve the process. For example, if it is found that a bolted electrical connection sometimes loosens during the burn-in process causing failures, then we would perhaps add an air-driven torque stop wrench to the process to always be used to install the offending hardware to the exact required torque each and every time.

This will cause the number of burn-in failures to reduce and perhaps allow the reduction of the required burn-in time. This is only one simple example of how feedback can and must be used to modify the process, and drive the correction of faults to their point of origin. While this concept of continuous feedback is applicable in both manual and automated processes, its implementation is best effected in automated systems where the data feedback always takes place, not just the seemingly important data, but as much data as the manufacturing engineer requests of the process.

Manual versus Automated

Manual processes are used where volumes are low enough that automated processes cannot be justified. Where volumes or other criteria justify it, automation can increase quality and quantities if properly implemented. Quality does not happen by accident, it is carefully planned for and designed in, both in automated and manual processes. Product volumes and direct labor costs must be monitored as a small company grows, so that automation can be embraced when it is justified, and not before. Product mix is also an important factor for consideration. Some vendors with a high mix have used this to their advantage while still employing automation and small lot sizes. A minimum lot size of one is possible, and with a high mix of product of various sizes can greatly increase efficiency. Automation can start on a very small scale, automating those parts of the process that are most labor-intensive or in parts of the process where scrap rates are high and expensive. Slow transition from the full manual process to a fully automated process is usually the best plan. This way the changes in personnel will take place as required, and current manufacturing personnel will be able to learn new tasks and grow with the process.

Another area where automation can be cost effective is in power supply testing. The addition of an in-circuit tester (ICT) for use in diagnosing boards, modules, and even single board power supplies, and increase process throughput thereby reducing costs. With the addition of loop testing to low-cost full-function ICTs such as the GenRad in-circuit testers, some companies have increased first-time power-on testing yields to better than 99%. Fewer failures mean less diagnose direct labor and lower manufacturing costs for the product. Feedback from these testers to the manufacturing engineer responsible for the process can be used to increase the process yields even further. Companies manufacturing power supplies where the in-circuit testing and loop testing process has been installed have typically cut diagnostic technician direct labor costs in half or better. Like any good process, success will depend largely upon the manufacturing engineers implementation of changes and improvements suggested by the feedback data from the process. The process becomes a live entity, in that it is always changing and improving.

Make versus Buy

In making the decision between buying versus building your own test equipment there is one important factor to consider that most engineers and managers tend to forget. That factor is that most forget that their company is in business to make a product, and not to design testers. Any energy directed at building your own testers takes away design effort from designing product. Designing product, not testers, is the purpose of engineering. Even in larger companies where there are whole test engineering groups, the total cost of overhead of which these engineers are a part of, are a part of the total corporate, engineering costs. When those costs are diverted from designing new products or making product improvements to designing testers that could have been purchased outright, the future of the corporation in new products may be put at risk. The reason that so many companies design and build their own testers is that they *always* underestimate the scope, size, length, and costs of the project. In the end analysis, the cost to the company is always greater, when a few test systems are built, than purchasing a comparable test system from a company that manufactures them in volume. In the case of buying a test system, the risk of losing the support of the test system's manufacturer is much smaller than the risk of losing the key designer of an internally built test system or of that person's time being consumed by some other more important project. So when is it right to build? Building your own testers is logically correct when for some reason you cannot buy the tester that you need for your product. Usually, however, you can easily purchase vendor modified equipment to suit your requirements. If, for example, your product has an unusual load requirement, a custom or modified load could be purchased from your test equipment vendor. If, however, your new design is of such a proprietary nature that you really wish to keep your new technology or unique application of technology under wraps until after you are in full production, designing your own test fixtures and even your own testers may be the correct decision.

In all instances, the best decision can be made when all of the facts are available. In considering your own design, you would also compare it with systems readily available off the shelf, even if those systems were customized for your particular application. To keep systems and options available off the shelf that have the required capabilities means talking to vendors and helping to keep them on the leading edge of technology by providing them with strategic information on the direction of the power supply test marketplace. Most power supply tester vendors have relatively small research and development (R&D) budgets and it is crucial that it be spent as wisely as possible. Providing white papers or position statements to these vendors on the directions that power supply manufacturing and test are going in will help ensure that you will be able to buy the equipment you need for your process, when you need it.

Another factor in making the choice of make versus buy is the universal nature of purchased test systems versus the typically dedicated nature of custom-built test systems. A universal tester will be applicable for many products already in production and future products not yet designed. This is largely due to the modular nature of power supply test systems. At some time in the future more loads may be added, different sized loads may be added, or new or different instrumentation capabilities may be added. This will spread the capital costs across many products for a number of years and greatly increase the return on the capital investment that the company has made.

Choosing a Tester

The process of choosing a tester can be a easy task if it is performed with a well thought out process. The checklist for choosing a tester (Table 15-1) typically includes the following steps:

Table 15-1. Checklist for Choosing a Tester

	Define Team Members
	Define Product Scope
	Define Specifications
	Define Vendor List
	Gather Vendor Data
	Evaluate Vendor Data
	Make Decision
	Publish Decision

Defining Team Members

It is recommended that the team be defined as consisting of persons from at least the following areas: Marketing, Sales, Design Engineering, R&D Engineering, Purchasing, Finance, Quality or Inspection, Manufacturing, Manufacturing Engineering, Service, Applications Engineering, Facilities, Corporate Management, and any other applicable group within your organization that will be affected by the purchase or equipment addition to the plant. Remember that if you don't get everyone involved from the very beginning, some may come forth at the last minute and delay the project while they are brought up to date and sold on the idea. Those who helped in making the right decision will support that decision that they had a part in making.

Some of the team members who may be very busy, may ask to be kept advised of the progress of the project, and they will then be invaluable allies if some resistance is encountered. Let people put in the level of effort that they feel comfortable, but spread the tasks around as much as possible.

Defining Product Scope

In choosing a tester, it is very important to understand the product mix that will be expected to use the tester not only today, but in the foreseeable future. You will need to examine the sales projections into the foreseeable future, today's product mix, the product mix for the next 5–7 years, product volumes for today and the next 5–7 years, and any special customer requirements that can be foreseen or projected. The further that you project out in time, the less certain those projections are. But these projections are useful in considering the direction that the company is going in so that the tester purchased gives the company to proceed in that chosen direction.

Defining Specifications

A first step is defining a set of **MUST HAVE** specifications. This will allow you to easily narrow down the field of vendors that you will be choosing from, making the job much easier. Additionally, you will want to also specify a list of **WANTED** specifications and finally a list of **NICE TO HAVE** specifications. These nice to have specifications are valuable because it is sometimes easy to get a capability designed in at little or no cost, especially when ordering customized equipment. If that specification were to be added later, it could be expensive or impossible to facilitate. An example of these three categories is a follows:

INPUT AC POWER SOURCE

- MUST HAVE   100–130 V
- WANT   90–250 V
- NICE TO HAVE   80–264 V

Any system not meeting the absolute list of must have's will be automatically eliminated.

Defining Vendors

Defining a list of vendors requires the examination of several characteristics of the vendor companies. These characteristics include: Test System Availability, Risks, Vendors Financial Stability, Vendor History, Support Availability, Training Availability, Maintenance, Upgradeability, Replication over time, Documentation Availability, Total Cost, And Projected Obsolescence.

Caution should be exercised when considering new ventures or companies just starting in a new market area. Companies have lost millions of dollars by believing that capabilities would be available just as a vendor promised to deliver and then didn't. A company entering a new market area may discover that it does not wish to continue in that market, leaving those companies who bet on its success in this new venture feeling the loss. Experience and a good track record greatly reduce the risk.

Gathering Vendor Data

Vendor data can be gathered from many sources, including: the vendor, customers, Dunn & Bradstreet, the Better Business Bureau, previous' years annual reports, and published articles in magazines and newspapers. Purchasing personnel can be of great assistance in this task.

Evaluating Vendor Data

In-Circuit Tester Vendors

To assist in the evaluation of In-circuit Testers a matrix of In-circuit Tester vendor evaluations is suggested. At the end of this chapter you will find a sample of this matrix form. On the first line is the list of vendors that have been chosen for evaluation. These make up column headings. In the sample form five vendors are listed at random; they are GenRad, Hewlett Packard, Zehntel, and others. Insert the names of the companies that you wish to consider in these spaces.

Each row then lists the item to be listed for each vendor. They are: **Model, Cost, Operation Costs, Startup Costs, In-Plant Service, Mail-In Repair Service, Training And Location, CPU Type, Networking, and Modularity. Also, Instrumentation which includes: Resistance Measurement, Capacitance Measurement, Inductance Measurement, Analog IC Testing, Semiconductor Testing, Guarding, Loop Testing, Additional Bias Supplies, and any Special Test needs. Additionally included are: Expandability, Years In Business, Financial Position, Technology, Software, and Automated Test Generation. Finally, included are: Total Population, Shipments This Year, Shipments Last Year, and Shipments the Preceding Year.** Just below the matrix are a list of the other vendors that were considered.

See sample at the end of this chapter.

Power Supply Tester Vendors

To assist in the evaluation of power supply testers, a matrix of power supply tester vendor evaluations is suggested. At the end of this chapter you will find a sample of this matrix form. On the first line is the list of vendors that have been chosen for evaluation. These make up the column headings. In the sample form vendors are listed at random; they are Autotest, Schaffner, in-house, and N.H. Research. The column titled in-house is listed so you may consider your own design against that of the others in the matrix. You may decide to choose a different set of vendors to suit your needs as this list may not be up to date. You can reach them at http://www.autotest.com, http://www.NHRESEARCH.com, and http://www.schaffner.com.

Each row then lists the item to be listed for each vendor. They are: **Model Number, Average Cost, Fixture Costs Plus Development Costs, Start-up Costs, Field Service in Each Area, Mail-in Repair, In-Factory Training, On-**

Site Training, CPU Type, Modularity Rating 0-9, Population, Previous Years Shipments, AC Source Rating 0-9, DC Loads Rating 0-9, Expandability Rating 0-9, Technology Rating 0-9, Automated Test Generation Rating 0-9, System BURN-IN, Years in Business, D&B Rating, and Overall Rating. At the bottom of the chart is the line listing Other Vendors Investigated.

Refer to Tables 15-1 and 15-2.

Making the Decision

The business factors to be considered when making the decision include product reliability, cost of goods sold, inventory levels, sales projections, current production test processes, long range test process strategy, planned process changes, current test processes, new processes, projected production costs, current production processes, testers, tester software, military test processes, cost of capital, volumes, product mix, quality, and customer requirements.

Each needs to be examined in how they directly affect the test process. Each of these factors need to be looked at and understood, in light of how they affect the business of today and tomorrow. Remember to consider that the product mix, volumes, costs, and technologies will always be changing. Your selection will determine how your company will keep a step ahead of competition, plotting a course into the future.

Once the data is gathered, the committee that you have established will meet to do the actual ratings and reach a decision. Remember that as you lead the committee toward a decision, that each member's viewpoint is very important and cannot be dismissed lightly. Resolve any problems outside of the meeting environment. This process can be a time consuming. Throughout the process, remember that some vendors will publish specifications that they cannot yet meet, but may hope to. It is always good to check certain system specifications on each system with a simple test so that you will have confidence in the specifications given where it is warranted. Where the call is close, it is often wise to favor the vendor having the most experience and the best reputation.

Publishing the Decision

When the decision has been reached, the committee must publish a formal memo signed by each member giving the decision. This is important for a number of reasons, but especially to show those in management and those who only observed that the task was started and now is completed. This will make it a successful task in their eyes. The next time it will be easier to get assistance with such a project.

Try to keep is short but with enough data to back up the decision, referring to the raw data as the source. Make all of the data available to anyone inside the company, but make sure that none of the evaluation particulars are given to any of the vendors, winners or losers. Each entrant need only be told that he or she either was or was not selected. Nothing else is required. Because the evaluation is done in light of your corporate testing requirements, you have no need to share that sensitive information with outsiders or vendors.

Tests

WARNING

TESTS IN THIS SECTION ARE NOT IN ANY LOGICAL SEQUENCE. DO NOT INFER ANY ORDER OF TESTING FROM THE ORDER THAT THESE TESTS ARE LISTED OR DISCUSSED. THE INFORMATION PRESENTED IN THIS HANDBOOK MAY BE SUPERSEDED BY LOCAL STANDARDS OR ORDINANCE. **ALWAYS CONSULT LOCAL STANDARDS OR ORDINANCES FIRST.**

Screening Tests

Screening tests are most easily performed during 'loop testing' on a low-cost in-circuit tester such as the GenRad GR2280i, when such a tester is available. (The software in the GenRad GR2280i makes it the most complete, well thought-out, user friendly, and productive system for power supplies that this author has seen.)

Output Backpower Test

The purpose of the output backpower test is to check the outputs for shorts prior to power being applied to the power supply under test. This test is performed by applying a DC voltage that is current limited, to individual outputs with the input power turned off, and monitoring the value of the current drawn by that output. This test is especially useful prior to applying AC power to a power supply that has outputs without current limiting, or where the current limiting function may not be working.

Bias Backpower Test

The purpose of the bias backpower test is to verify that a short or open condition doesn't exist on the bias supplies. This test is performed by applying a DC voltage which is current limited to the bias supplies in the unit under test and measuring the actual current for comparison to the expected result. Loss of a bias supply with the application of full input power can be catastrophic to the unit under test.

Reference Voltage Checks

Most power supplies contain reference voltages which are used to set the output values. The purpose of this test then is to check each reference voltage in the power supply under test prior to power being applied to the supply. A missing or incorrect reference voltage can cause outputs to vary well beyond design limits. This test is accomplished by supplying the required bias supplies that are used to create the reference voltages and then measuring the reference voltages at the appropriate points.

Overvoltage Trip Point Test

Any power supply with an overvoltage protection circuit will require the testing of each outputs overvoltage trip point prior to power being applied. The overvoltage trip point can be tested by applying current limited back-power to an output of interest and monitor the point at which an indication of the trip point being reached. For some power supplies the indication is a signal indicating that the overvoltage trip point has been reached, while for others it may be the fact that a "crowbar" circuit has tripped shorting the output under test. Whatever the indication for the unit under test is, it must be measured and the voltage at which it was noted is then the overvoltage trip point.

Overvoltage Trip Point Recovery

Because some power supply designs allow for the recovery from an overvoltage condition without going through a power off condition, it will sometimes be necessary to make sure that the outputs recover from having an overvoltage applied to the output within a specified set of conditions. The actual design specification for the overvoltage trip recovery will be used to create the required test.

Primary Overcurrent Functionality Test

The purpose of this supply-dependent test is to make sure that the primary overcurrent or overpower detection and/or protection circuitry is functioning prior to power being supplied. The format of this test is determined by the circuitry to be tested.

PWM Oscillator Frequency Test

This is a test of the Pulse Width Modulator (PWM) oscillator frequency to determine that it is within the expected range of values. This typically involves a gross check to determine that the frequency deviation is not double or half the expected value or greater, prior to the application of full power.

PWM Functionality Test

The PWM functionality test verifies the functionality of the pulse width modulation capabilities of the PWM prior to power-up. In fixed pulse width variable frequency systems, this test checks the variation in the off time of the PWM.

External Sync Check

On power supplies where a provision has been made for an external sync signal, the external sync test verifies that external sync input is working.

External Frequency Shift Test

On power supplies where a provision has been made for an external frequency shift signal, the external frequency shift test verifies that the external frequency shift input is operational prior to power being applied.

Input Ramp-Up at Minimum Load Test

The input ramp-up at minimum load test is designed to check that a short doesn't exist in the input section of the supply under test prior to power being applied. A low voltage AC source is applied to the input and the current is monitored. If the current exceeds a preset limit, a shorted condition exists in the input circuitry, and needs to be diagnosed and repaired before any further testing.

Short Test, Main Switch Transistor

Catastrophic damage can occur to a power supply where the main switching circuitry has a short when power is applied. Therefore this test determines that no such short exists. One way to be certain that a short doesn't exist on main switching transistors prior to power being applied to the supply is to apply bias power and then a current-limited 10 or 20 V source to each of the transistor switches in turn. A measure of the AC voltage across that switch will tend to indicate that the switch is operational. Additionally, it is a good practice to place bias supplies on the power supply under test and then ramp up the DC supply to the main switches, while monitoring the current for an indication of a problem. Most switching power supplies will, if not loaded at all, produce a regulated output with less than 100 VDC input. Many will produce full output voltage with 40–60 VDC of input. Results will vary with the design topology.

Experience has shown that if this type of extremely low power test is performed on an ICT that the following test step will have much higher yields, typically over 99%. This is due to the precipitation of faults at first power application being discovered in an environment where the unit under test (UUT) is protected from high voltage and high current surges and the resulting catastrophic damage that can occur.

AC Input Tests

These tests include:

Soft Start

On power supplies that contain some type of softstart functionality, this test is designed to check that circuitry. Since there are many methods of achieving softstart, it is not possible to describe a single test that would work in all cases. Some softstart circuits reset in microseconds while others may take many seconds, the timing is dependent upon the type of circuits and the components used to facilitate the soft-start function. In creating a test for this function, the power supply specifications and circuitry must first be examined carefully.

Inrush Current

The inrush current test is designed to test the value of the surge current at tumon in the input of the supply under test. At a specified point in the AC input cycle the power is turned on. Typically, this point is selected to be at or near the phase where the worst case or maximum value current will be realized. As the AC source is applied to the input, the current is monitored with instrumentation capable of obtaining the instantaneous peak value of current. If the current exceeds a preset limit, then a fault condition exists in the input circuitry, and needs to be diagnosed and repaired.

Power Factor

The power factor is the ratio of true to apparent power expressed as a decimal. In today's power supplies, power factor is more complex than just a phase difference between the voltage and current waveforms. Distorted current and voltage waveforms create a power factor that cannot be easily measured with a simple oscilloscope. Some power supplies employ power factor correction (PFC) circuitry to reduce the amount of distortion in the current waveform. Such power supplies commonly use the Micro Linear ML4821 or ML4822 Power Factor Correction Control IC. Additionally, Micro Linear produces Combo chips that contain both PFC and PWM in a single monolithic chip. This means one clock, and one frequency to worry about for FCC testing. Power supplies employing some form of power factor correction will need to have the power factor of the input power check, at least at new product start-up, later this test may be dropped once it has been demonstrated that it is no longer required. For more information see: http://www.microlinear.com.

Static Line Regulation

A check of the output of the power supply under test while the output load is held constant, while the input is varied form nominal to minimum and maximum values. The output is measured for variations due to the different line values.

Dynamic Line Regulation

The output of the power supply under test is measured while the input is varied from the specified minimum and maximum values and the load is at specified minimum load and maximum load conditions. The change and rate of change of the input must be taken from the power supply specifications. This test also provides us with the line transient response data.

Start-Up Time

Start-up time is the time it takes after the application of input power, at full load and low line conditions, for all outputs to become stable within the specified limits.

Hold-Up Time

Hold-up time is the time, at full load and low line conditions, that a specified output or outputs, will remain within the specified limits, after the removal if input power.

Dropout Time

Dropout time is the maximum time that the input power can be removed and reapplied, at full-load and low-line conditions, and maintain regulation within specification on all outputs.

High Line Minimum Load Regulation Check

High-line minimum-load regulation check is a measurement of output voltages at high-line and minimum load conditions. Specify the switch position for 115–230 V power supplies, where applicable.

AC Input Current (RMS)

AC input current is a measurement of the AC RMS input current at the full load worst case condition of either low line or high line voltage.

Efficiency

Efficiency is a measure of the power consumed in the process of power conversion. For AC/DC and AC/AC converters, the measurements process requires that careful attention be paid to the RMS measurement of both voltage and current, as well as the phase relationship between them. The total AC power is then calculated from the measured data. Instruments are available that can make this measurement directly. Most all automated power supply testers have this capability as well.

DC Input Tests

For both DC to DC converters and DC to AC inverters we test the units with DC inputs. The following tests are some you may wish to consider in testing these types of power supplies.

Softstart Test

On those power supplies that contain some type of softstart functionality, this test is designed to check that circuitry. Since there are many methods of achieving softstart, it is not possible to describe a single test that would work in all cases. Some softstart circuits reset in microseconds while others may take many seconds. The timing is dependent upon the type of circuits and the components used to facilitate the softstart function. In creating a test for this function, first examine the power supply specifications and circuitry carefully.

Inrush Current Test

The inrush current test is designed to test the value of the surge current at turn-on in the input of the supply under test. The power is turned on the DC source is applied to the input and the current is monitored with instrumentation capable of obtaining the instantaneous peak value of current. If the current exceeds a preset limit, a fault condition exists in the input circuitry, and needs to be diagnosed and the design modified or repaired.

Static Line Regulation Test

The output of a power supply under test is measured while the input is held constant. The output is measured at minimum load and maximum load conditions.

Dynamic Line Regulation Test

The output of the power supply under test is measured while the input is varied from the specified minimum and maximum values and the load is at minimum load and maximum load conditions. The change and rate of change must be taken from the power supply specifications.

Start-Up Time Test

Start-up time is the time it takes at full-load and low-line conditions for all outputs to become stable within the specified limits after the application of power.

Hold-Up Time Test

Hold-up time is the time, at full-load and low-line conditions, that a specified output or outputs will remain within the specified limits after the removal of input power.

Dropout Time Test

Dropout time is the maximum time that the input power can be removed and reapplied, at full load and low line conditions, and maintain regulation within specifications on all outputs.

Efficiency

Efficiency is a measure of the power consumed in the process of power conversion. For DC-DC converters, it is simply the Watts Out divided by the Watts In, multiplied by 100%.

Control Input Tests

Many power supplies have control inputs that perform various functions as noted in the power supply specifications. Each input specification and circuit will need to be examined carefully to gather the data needed to write the correct test.

DC Output Tests

The following tests provide you with a selection of those tests that you may wish to perform on the DC outputs of your power supplies.

Static Load Regulation Test

The output of a power supply under test is measured while the load is held constant. The output is measured at minimum line and maximum line conditions.

Dynamic Load Regulation Test

The output of the power supply under test is measured while the input is held constant at the specified nominal value and the load is dynamically switched between specified load conditions. Ideally, the power supply under test will handle switching between minimum load and maximum load conditions. In reality, few power supplies can tolerate dynamic switching between minimum-load and maximum-load conditions. Factors that can greatly affect dynamic load regulation tests include the power supply under test, the rate of transition, and the broadband impedance of the interconnecting cable. In some power supplies the "crowbar" or overvoltage circuit will trip due to the inductive nature of the interconnecting cable at fast switching transitions. Modem electronic loads can typically only switch at a rate up to about 20 A/µS. The length of cable between the unit under test and the load is very important. If the cable is only 1 or 2 ft long the parasitic inductance value may be low enough to not cause any problems. If, however, the interconnecting cable is 6 ft or greater, in length, the parasitic inductance value may be sufficient to cause improper operation and may even damage the unit under test in certain cases.

To solve the problem of inductive cabling, a low impedance cable, matching the impedance of the load, should be used. This cable is however not generally available at this writing, however the concept has been patented.

To determine the correct rate of switching in amperes per microsecond, examine the specifications of the power supply in question, the actual load that it will be connected to, and the type and length of interconnection cable that is specified. Some power supplies can perform well with rates of 20 A/µS from 20 to 80% of rated load, while others require much slower slew rates. Each design has been created specifically with an application, or range of applications, in mind, and therefore has different specifications. During the dynamic load regulation test the transient response time can be measured.

PARD or Ripple and Noise Test

PARD stands for the Periodic and Random Deviation of the DC output voltage from its average value, over a specified bandwidth and with all other parameters maintained constant. This test is made at nominal line input voltage and full rated voltage and current. The accepted bandwidth, unless otherwise, specified, is from 20 Hz to 20 MHz. Variations of less than the specified lower frequency limit are taken as drift. The measurement needs to be made both with a RMS reading and a peak to peak reading. Ensure that the measurements are made with a setup that does not create ground loops giving faulty results. To check the setup, turn off the power supply under test and check for a faulty measurement.

Ripple

The RMS reading is primarily indicative of the low-frequency components of ripple voltage at the line frequency, the switching frequency, or some harmonic or product of the two. This reading should be made with a capacitive coupled connection to eliminate DC drift from the measurement.

Noise

The peak-to-peak measurement is the measurement of those small noise spikes especially found in switching power supplies, and the typical shot noise present in electronic circuits. While noise spikes do not appreciably add to the RMS value, they can cause faulty operation of the product. This is why noise is measured as a peak-to-peak value using a good oscilloscope having the desired bandwidth

Short Term Drift

Short term drift is measured after specified warmup period, typically 30 min. It is measured with all conditions controlled and steady, including input voltage, line frequency, ambient temperature, and load current. Short-term drift is specified as variations in the output voltage of less than 20 Hz and shorter than 30 min. The test is usually 30 min long but may be specified otherwise.

Long Term Drift

Similarly, long-term drift is measured after specified warmup period, typically 30 min. It is measured with all conditions controlled and steady, including input voltage, line frequency, the ambient temperature, and load current. Long-term drift is specified as variations in the output voltage of less than 20 Hz. The test is usually 8 hr long but may be specified otherwise.

Low-Line Full-Load Regulation

The low-line full-load regulation check is a test used by some as a check of the value of the bulk capacitor. The results of this test are important to those repairing product give an indication of the value and condition, of the bulk storage capacitor. Aluminum electrolytic capacitors age, their effective value decreases as they dry out.

Overcurrent Tests

These tests check the maximum overcurrent trip points of the Power Supply under test.

Current Limit

This type of power supply limits the current to some maximum value allowing the output voltage to reduce toward zero, thus preventing the output current exceeding the maximum value. This type of current limiting scheme is not used often in linear regulators as it often creates intolerable output device power dissipation in linear regulators. This scheme is sometimes used switching regulators.

Current Foldback

Where device and or heatsink limitations exist under shorted conditions, foldback current limiting becomes the superior alternative to standard current limiting. In the test for foldback current limiting, the maximum current available at full load must be measured as well as the lower value of short-circuit current.

Overcurrent

Some overcurrent schemes will latch-up when activated and must be manually reset once the shorted condition is removed. This test can be difficult to perform without test equipment that can measure the peak current value reached because as soon as a sorted or overcurrent condition is reached the power is removed.

Overcurrent Recovery Test

This is simply a test to determine is the overcurrent recovery acts as expected in any of the aforementioned overcurrent protection schemes.

Short Circuit Check

This test checks the short current capabilities of the power supply outputs.

Short Circuit Recovery Check

This test checks to see if the power supply will regulate after a short has been applied to the output and then removed.

Timing Tests

Timing tests check the timing relationship between AC Low and DC Low, Power OK, and other circuits as required by the specification. It is common to specify the order in which a power supplies outputs will come up and go down.

Switch Test 120 V

Where there is a 120/240 V line selection switch, this test checks that the Power Supply will regulate in both switch positions.

LED Output OK Checks

The purpose of the light emitting diode (LED) output OK check is to test that all LEDs or other visual outputs are functioning correctly.

Overvoltage

There are two meanings to overvoltage testing that we need to examine. The first is related to overvoltage protection or "crowbar" circuitry where a voltage is applied to the output terminals, from a current limited source, until the overvoltage protection circuitry reacts. The voltage at which the overvoltage protection circuitry reacts is the value that is of interest. The second is related to the input circuitry and its ability to withstand an overvoltage without damage.

Isolation

Power supplies used in critical military, aerospace, medical, and communications systems may be designed to provide a certain level of isolation between the load side and source side of the power supply. Isolation is usually related in decibels at a specified frequency range.

Cross-Regulation

On multiple-output power supplies, cross-regulation is defined as the effect that a load variation in one output has on the other. One output is provided with a constant load while the other is dynamically switched between two load points. While the actual load variation is usually determined by the application, a 20 to 80% load swing is typical. Of equal importance is the rate of change in the load current. 20 A/μS is about as fast as commercial loads will go at this writing. Depending upon the application, rates of 1, 5, 10, and 20 A/μS are common.

Short Circuit

WARNING

Never short a power supply output that has not been designed with some type of short protection. **NEVER** use a screwdriver or other tool to short an output!

Short testing is necessary to determine how much current the power supply under test will deliver into a short. Most power supplies with overcurrent protection have the maximum current under shorted conditions, that is, zero volts of output, specified.

AC Output Tests

Those power supplies that have an AC output to drive fans or some other purpose will require AC testing. The following are some tests that you may wish to consider.

Static Load Regulation

Static load regulation is a test of the output voltage (RMS) under the nominal, as well as the minimum and maximum specified input voltage.

Dynamic Load Regulation

Three tests are used to test a dynamically varying AC load. The first test is done at a rate considerably less than the output frequency, the second is performed just above the output power frequency, and the third at a rate considerably than the output power frequency. The highest frequency of the three tests should be a square wave, while the others may be either sine wave or square wave.

Frequency

Frequency is simply a measurement of the output frequency under various conditions of line and load as specified.

Waveform

Some AC outputs may be designed to be square waves or other waveforms depending upon the application.

Spectral Purity

For sine wave output AC sources, spectral purity is a measure of harmonic content of the AC power output, usually at minimum load, nominal load, and maximum load conditions.

Ripple and Noise

Ripple

Ripple is the content in the output of the line frequency. In non-sinusoidal outputs, undesirable harmonics of the output power frequency may also be called ripple.

Noise

Noise is the random deviation in the AC output not specifically related to the line frequency or the output power frequency.

Ancillary Testing

Many power supplies have ancillary outputs of some kind or other. Let's examine a few common ones.

ACOK

ACOK is a measure of the input line power. This signal is typically used in computer power supplies to warn a computer central processing unit (CPU) that a power loss in about to happen.

DCOK

DCOK is a signal sent by the power supply to indicate that all outputs, or in some cases, some outputs are grossly within specification.

Real-Time Line Clock

A real-time line clock signal is a logic signal created by the power supply from the AC line frequency. Filtering is commonly used to remove transient signals from falsely incrementing the line frequency count.

HIPOT Testing

HIPOT testing is well defined by the regulatory agencies around the world.

ALWAYS EXERCISE EXTREME CARE WHEN PERFORMING ANY HI-POT TESTING. DANGEROUS VOLTAGES AND CURRENTS ARE PRESENT.

Table 15-2. Matrix of In-Circuit Testers

Matrix of In-Circuit Testers		GENRAD	HP	TERADYNE	ZEHNTEL	OTHER
MODEL						
Machine Cost						
Operational Costs						
Start-Up Costs						
Field Service	In-Plant					
	Mail-In					
Training & Location						
CPU Type						
Operating System						
Networking						
Modularity						
Instrumentation						
S	Measure Resistance					
P	Measure Capacitance					
E	Measure Inductance					
C	Measure Analog Ics					
S	Meas Semiconductor					
	Guarding					
	Special Test Needs					
	LOOP Testing					
Expandability of Tester						
Years in Business						
Total Tester Population						
Technology						
Financial Position						
Software & Auto Test Generation						
Vendor Support						

Table 15-3. Matrix of Power Supply Testers

Matrix of Power Supply Testers				
Vendors Name	Autotest	Schaffner	IN-HOUSE	N.H. Research
MODEL NUMBER				
COST OF TESTER				
FIXTURE + DEVELOPMENT				
START-UP COST				
FIELD SERVICE				
In Each Area				
Mail-In Repair				
In-Factory Training				
On-Site Training				
CPU Type				
Modularity				
Population				
Previous Years Shipments				
AC Source				
DC Loads				
Expandability				
Technology				
Automated Test Generation				
System Burn-In				
Years in Business				
Dunn & Bradstreet Rating				
Overall Rating				
OTHER VENDORS INVESTIGATED: _____				

16

Burn-In, Quality, and Reliability Testing

We will start by examining the concept of burn-in as a quality and reliability test tool. The one key concept to keep in mind throughout this chapter is that a modern manufacturing process is not a fixed entity, but a live and ever changing creation. For it to be successful, it must by definition, be always changing and improving. The concept of constant and never-ending improvement must be a filter through which we view the issues surrounding QUALITY, RELIABILITY, and BURN-IN.

Burn-In, the Tool

Creating a universally acceptable concept of the burn-in process and the terms used in describing the burn-in process is a nearly impossible task. To start with, just about everyone has a different concept of the problem, the solution to the problem, or even of the need for burn-in at all. Perhaps a better way to approach this problem is by examining power supply quality itself. The first step will be in defining the terms used in discussing the process, for common terminology is required to have a better understanding of the many different facets of the process.

The second step will be to examine the processes used for power supply burn-in. The actual processes used will vary by:

- Class of product (markets)
- Product volumes
- Company size
- Applications

Markets may include:

- Aerospace
- Commercial
- Computer
- Consumer
- Industrial
- Instrumentation
- Research
- Medical
- Military
- Telecommunications

Within each of the classes, various levels are generally proscribed that may include the full gamut from the minimum of no burn-in to the maximum, which may include such processes as:

- Power Cycling
- Elevated Temperature (heated)
- Elevated Ambient (self heated)
- Reduced Temperature (cooled)
- Temperature Cycling
- Thermal Shock
- Cold Soaking
- Vibration
- Shock
- Humidity
- Radiation
- Altitude / Pressure
- Powerline Perturbations

The level of burn-in required may be specified in such a manner that actual burn-in time may be reduced or eliminated, by a controlled process that will continue to ensure the quality of the product. Process specifications

Table 16-1. Possible Burn-In Option Selection Chart

Markets	Pwr Cycle	Hot	Elevated ambient	Cold	Temp. cycling	Thermal shock	Cold soaking	Vibration	Shock	Humidity	Radiation	Altitude	Power pertabat
Aerospace	•	•		•	•	•	•	•	•	•	•	•	•
Commercial	•	•											
Computer	•	•			•								
Industrial	•	•		•	•								•
Instrument	•	•		•	•		•						•
Research	•	•					•						•
Medical	•	•		•	•		•					•	•
Military	•												
Telecom	•	•			•	•	•	•	•	•	•	•	•
Other				•									

must include criteria for the evaluation of both the actual burn-in and the process. This will aid in the observation and documentation of design and production anomalies, as well as pointing out concealed problems with components. See Table 16-1 for Possible Burn-In Option Selection Chart.

While in the fundamental analysis burn-in is generally used to effect product reliability, in too many cases, the operation is performed without a clear understanding of the goals of the process. All too often, it may be performed at typical conditions, and no real stressing of the product occurs. Let's first define the process phases for burn-in:

- Phase One: Pilot Production
- Phase Two: Production Startup
- Phase Three: Process Improvement
- Phase Four: Production Maturation
- Phase Five: Ongoing Reliability Assurance

Lets examine each of these process phases in more detail:

1. *Phase One: **Pilot Production:*** Burn-in for pilot production must be the most strenuous because it is testing not only a new product, but usually a new or different process as well. Due to the low volumes, longer burn-in times are more easily achieved than they would be for higher volumes. But how much is enough, or too much?

2. *Phase Two: **Production Start-up:*** At production start up the process normally starts to settle out a bit, process and product anomalies are discovered and the root causes attacked. Sometimes additional tests or process steps are added with the goal of driving fault detection back to the point of fault generation. This is the least expensive place to find a fault, right where in occurred.

3. *Phase Three: **Process Improvement:*** This is a continuous process of improving as the process continues through ramp-up, creating an ever more flexible/mature process that produces quality products at reasonable costs. The key to success is the constant timely feedback received from the process, and quickly acted upon, which allows the fine tuning of the process and the product to approach perfection.

4. *Phase Four: **Production Maturation:*** As the process and product mature, process stability increases, the level of attention needed to tweak the process and/or product decreases to the point where only occasional attention is need as maturity is reached. Occasional perturbations in product quality may dictate temporary changes in the process or process parameters until the source of the problem is located and fixed. Once stability is reachieved and assured, the process will settle into the steady state mode again.

5. *Phase Five: **Ongoing Reliability Assurance:*** As a check of the process, some percentage of the product produced by the process is routed through the additional burn-in and testing. As products and processes mature, burn-in times may be reduced or even eliminated. The Ongoing Reliability Assurance program provides a longer burn-in on a sample of the product, and perhaps additional testing to test the process. This is our sanity check that using the process that is doing reduced burn-in still does produce quality product.

Goals

The types of measurements performed are determined by the types of units under test, the failure modes of concern, the precision of expected results, the accuracy of measurement tools, the costs involved, the design tolerances, and importance of parameters, among other factors.

Process Specifications

The development of the process must have one key goal as the central focus, that is, **"The development of a process where timely feedback causes continual improvement of the process."** Every process will be different, and will have a different set of specifications. Process development is not a one person job, but requires a team of individuals from many disciplines. Your first task is to form a team with members from every area of the company, and then encourage them to share in your vision of the development of a process where timely feedback causes **CONTINUAL IMPROVEMENT** of the process. This is the only real way to reduce product costs and to improve yields and therefore quality.

Feedback

It is possible to use the test and failure data to statistically determine the shape and length of the initial part of the bathtub curve. Some situations are extended to life-testing, where the intent is to find the statistical life expectancy of a particular type of component or assembly, and thus the shape and starting point of the final part of the curve. In addition, burn-in may be used to accelerate early component failures, so that surviving components may be reasonably expected to be in the flat portion of their life-cycle curve. In the statistical instances, sampling is done to provide enough data for analysis. In the final case, 100% testing is required to adequately isolate early component failures. The implementation of burn-in for statistical purposes may be a temporary process step which is eliminated when the statistical maturity of a process and the components is assured.

Depending upon the product and the components used, and upon factors such as typical use environment and penalties of failure, burn-in may be implemented at one or more places in a design and/or manufacturing process. Military or aerospace applications demand that all components be in their "normal life" portion of the bathtub curve, and that they be highly predictable in terms of minimum lifetime. This regimen is necessary because military or aerospace products may be used in severely harsh environments, and or in situations where human life depends directly upon their operation. In the case of satellites and space probes, repair cannot be done after launch, so reliability is paramount.

Commercial products have different burn-in requirements, because commercial or assembly is not potentially life-threatening, and most products are in situations where repair may be effected. As would be expected, new products require more burn-in for analysis than traditional products. Very mature processes often do no burn-in for analysis, reserving the burn-in process step for precipitation of early failures. In the case where all components are purchased as "preburned-in" by the manufacturer, the burn-in step may be eliminated entirely from the mature production process flow, when all process faults have been eliminated. Countering this trend is the desire for increased quality assurance, confirmed by continuing burn-in, which pays back in reduced failure during warranty and perhaps allows a lengthened warranty period.

The costs of burn-in stem from equipment purchases, floor space required, burn-in station maintenance, some reduction of process throughput, labor to perform burn-in, and skilled analysis of results. In products with fairly reasonable volumes, these costs are generally insignificant. At low production volumes, burn-in becomes a considerable portion of production costs. The costs associated with burn-in may be plotted as an inverse exponential function with respect to volume or time. This is due to the same learning curve effect mentioned above.

Burn-in is most effective when process feedback is applied. The information to correct a process or component deficiency will be best used if timely analysis is done. By proper analysis, the process immaturity factor may be controlled, and the process quality will become understood and statistically predictable. Burn-in is a reliability tool, and as such, has the potential to monitor and enforce product reliability. The question is not whether or not to perform burn-in, but how, when, where, and to what extent.

As for the future; power products will become inevitably more complex and greater reliability will be demanded. The tools of burn-in, as it is practiced today, will not be adequate. More degrees of variation in line quality and load excursions will have to be accommodated. The volume of multioutput power supplies will necessarily increase.

A technique for solving the burn-in complexity demand for power products is available but unimplemented today. Since the early 1970s, a number of companies have been manufacturing electronic loads. And since, that same time, microprocessors have become inexpensive components. By using a dedicated microprocessor to monitor and control a burn-in station using electronic loads, a cost-effective means of performing complex power product burn-in may be achieved.

To perform the necessary control and monitoring, the system would have to control a number of electronic loads. These loads may be specifically designed into the burn-in controller to save space and cabling. An analog data acquisition section would facilitate measurements of the voltage and current impinging upon the loads. A method of electrically modulating the loads, to simulate typical use, would be required. While adjustment of the line voltage applied to the power product is very desirable, it may not be economically feasible. Controlled interruption of the line voltage is both desirable and very feasible. The burn-in controller should be capable of making decisions about the health of the unit being tested, so that a shutdown sequence can be entered upon catastrophic failure. A minimum of controls are required; including a start switch, an abort switch, and a display of unit under test status. A method of presetting the acceptable limits of measurements is required, but these parameters may be designed into the firmware of the microprocessor, or, in the case of a flexible unit, programmed from a terminal or optional front panel or remotely via the IEEE-488 bus.

Burn-in will be a part of power product design and manufacturing processes for all of the foreseeable future. The prudent use of burn-in where practical can only serve to improve product quality and process maturity.

Why Do Burn-In Anyway

One of the first questions that comes to mind as we start to examine and study the subject of burn-in for power supplies is "Why do we want to do burn-in, anyway?" The answer to that question may in part be the following: "To detect and remove process faults before they reach a customer." We manufacture products, containing power supplies, for use in the commercial world. The task of burn-in is not to locate and repair faults that will cause early failure of our products in a commercial business atmosphere. Burn-in is a test process step that is used to identify those faults in the process that identifies the manufacturing fouls caused by process inadequacies. The *real goal* of "burn-in" *is to provide timely feedback to correct the process* so that only good units are produced.

For most, if not all of our customers, it is not appropriate to test with the purpose of encouraging infant mortality, due to the negligible benefit, and the costs associated with performing the testing. Burn-in must be used to identify and correct process problems but need not be used to force infant mortality. The major differences lie in the length of time of burn-in and the severity of environmental stresses placed on the unit under test.

The following is a guideline to provide reasonable limits for creating burn-in process:

- Products will be burned-in, for example, 80 hr, starting at new product introduction, at a temperature that is 5° above the specified ambient for the product. This scheme includes prototypes and pilot production units.

- Products will be monitored during burn-in and all failures will be automatically logged for failure type and accrued burn-in time.

- Products will be power cycled at predetermined rates. The suggested rate is to cycle off for 20 S, then on for 5 min, off for 2 min, then on for a minimum of 2 hr. The goal is for the entire power supply to reach its thermal equilibrium.

- All failures will be repaired by manufacturing engineering to facilitate optimization of the process.
- All data on failed units will be correlated with previous process steps and corrective action taken. A process step or steps may have to be modified or added as required.
- All data on every failed unit will be correlated against accrued burn-in time. Burn-in time will be periodically adjusted, either upward or downward, as indicated by failure analysis. Suggested minimums for review are once every 100 units or 48 hr, whichever is greater.
- Where indicated, burn-in will be moved out of process, and performed on a sample basis only, as long as sample burn-in has results consistent with projected data.

Thus, a goal of burn-in is to reduce itself to the lowest appropriate level, when directed to do so by the process. Most products, with some improvement of the process, should be able to undergo a burn-in of 2 hr. Some products, at process maturity, should be able to tolerate burn-in being reduced to 30 min, with good yields. With the proper design and very tight process control, burn-in may be removed from the process, and be used to spot check the goodness of the process yield.

It is possible that testing on sophisticated burn-in equipment may be done in such a manner that it would incorporate process steps of: power supply functional testing, burn-in, and final power supply functional testing.

There are many different perceptions of burn-in. In fact, burn-in may not really be the appropriate name for the process step. Perhaps a name such as environmentally controlled stress testing better defines the process. One reason to change the name is to help remove the emotional responses that the term "burn-in" sometimes evokes.

The author encourages vendors to provide the hardware to achieve the testing indicated above. In fact, that testing may be done in such a manner that it replaced the current process steps of:

- Power Supply Functional Testing
- Burn-in
- Power Supply Functional Testing

A Power Supply Burn-In Standard

The types of measurements performed are determined by the types of units under test, the failure modes of concern, the precision of expected results, the accuracy of measurement tools, the costs involved, the design tolerances, and importance of parameters, among other factors.

Power supply burn-in is not well defined or well understood by everyone in the electronics industry. In fact, it is greatly misunderstood. Terms are not well defined across the legion of markets using power supplies. There are many types of Burn-in in use today. Even the goals of burn-in are not well defined in many cases.

The lack of unity in terms, goals, and uses of burn-in to date, has not created a well defined market for power supply burn-in equipment. Power supply burn-in processes lack any commonality further hindering the development of such a market.

The problem is therefore, to create a standard that will do the following.

- Define standard terms that will allow meaningful communication between members of the community talking about burn-in.

- Define burn-in by major markets due to the vast differences between these markets.

- For each major market, define an appropriate number of levels of power supply burn-in to meet the needs of that market.

- The process may consist of operation at artificially elevated and/or depressed temperatures, time, high or low input voltage, environmental tests, humidity, vibration, power cycling, atmospheric pressure, thermal presoak, and radiation, all dependent upon product specifications and requirements.

- Define a flexible burn-in processes that starts with new product introduction and continues through product process maturity and product maturity.

- Define the process for achieving zero burn-in while maintaining quality levels.

- Define the process of product quality and reliability measurement.

It is possible to use the test and failure data to statistically determine the shape and length of the initial part of the bathtub curve. Some situations are extended to life-testing, where the intent is to find the statistical life expectancy of a particular type of component or assembly, and thus the shape and starting point of the final part of the curve. In addition, burn-in may be used to accelerate early component failures, so that surviving components may be reasonably expected to be in the flat portion of their life cycle curve. In the statistical instances, sampling is done to provide enough data for analysis. In the final case, '100% testing' is required to isolate all early component failures. In many cases, the implementation of burn-in for statistical purposes may be a temporary process step, which is eliminated when the statistical maturity of a process and the components is assured.

Defining an ECST Tester

Depending upon the product and the components used, and upon factors such as typical use environment and penalties of failure, environmentally controlled stress testing (ECST or burn-in) may be implemented at one or more places in a design and or manufacturing process. Military or aerospace applications demand that all components be in their normal life portion of the bathtub curve, and that they be highly predictable in terms of minimum lifetime. This regiments necessary because military or aerospace products may be used in very harsh environments, and in some situations where human life depends directly upon their operation. In the case of satellites and space probes, repair cannot be done after launch, so reliability is paramount.

Commercial products have different burn-in or ECST requirements, because commercial or assembly is not potentially life-threatening, and most products are in situations where repair may be effected.

As would be expected, new products require more burn-in or ECST for analysis than traditional products. Very mature processes often do no burn-in or ECST for analysis, reserving the burn-in process step for precipitation of early failures. In the case where all, or nearly all of the components are purchased as 'pre-burned-in by the manufacturer', the burn-in or ECST step may be eliminated entirely from the mature production process flow, after all process faults have been eliminated. Countering this trend is the desire for increased quality assurance, confirmed by continuing burn-in or ECST, which pays back in reduced failure during warranty and perhaps allows a lengthened warranty period.

The costs of burn-in or ECST stem from equipment purchases, floor space required, burn-in station maintenance, a potential reduction of the process throughput, labor to perform burn-in or ECST, and skilled analysis of results. In products with reasonable volumes, these costs are generally insignificant. At low production volumes, burn-in or ECST becomes a considerable portion of production costs. The costs associated with burn-in may be plotted as an inverse exponential function with respect to volume or time. This is due to the same learning curve effect mentioned earlier.

Burn-in or ECST is most effective when process feedback is applied. The information to correct a process or component deficiency will be best used if timely analysis is done. By proper analysis, the process immaturity factor may be controlled, and the process quality will become understood and statistically predictable. Burn-in or ECST is a reliability tool, and as such, has the potential to monitor and enforce product reliability.

As for the future; power products will become inevitably more complex and greater reliability will be demanded. The tools of burn-in, as it is practiced today, will not be adequate. More degrees of variation in line quality and load excursions will have to be accommodated. The volume of multi-output power supplies will necessarily increase. The question will not be whether or not to perform burn-in or ECST, but how, when, where, and to what extent.

A technique for solving the burn-in or ECST complexity demand for power products is available but un-implemented today. Since the early 1970's, a number of companies have been manufacturing electronic loads. And since that same time, microprocessors have become inexpensive components. By using a dedicated microprocessor to monitor and control an ECST station using electronic loads, a cost effective means of performing complex power product ECST may be achieved.

To perform the necessary control and monitoring, the system would have to control a number of electronic loads. These loads may be specifically designed into the burn-in controller to save space and cabling. An analog data acquisition section would facilitate measurements of the voltage and current impinging upon the loads. A method of electrically modulating the loads, to simulate typical use, would be required. While adjustment of the line voltage applied to the power product is highly desirable, it may not be economically feasible. Controlled interruption of the line voltage is both desirable, and feasible. The ECST controller should be capable of making decisions about the health of the unit being tested, so that a shutdown sequence can be entered upon catastrophic failure. A minimum of controls are required, such as: a 'start' switch, an 'abort' switch, and a display of unit under test status. A method of presetting the acceptable limits of measurements is required, but these parameters may be designed into the firmware of the microprocessor, or, in the case of a flexible unit, programmed from a terminal, an optional front panel or remotely via the IEEE-488 bus.

ECST will be a part of power product design and manufacturing processes for the foreseeable future. The prudent use of ECST where practical can only serve to improve product quality and process maturity.

Burn-In Basics

For a long time, burn-in has been the term applied to a wide variety of processes and techniques that allow an electronic product to run under controlled environmental conditions in an effort to either determine or confirm statistical reliability estimates, that is, the product's goodness.

Burn-in can be used at any one of a number of points including; design verification, component verification, assembly verification and testing, and burn-in can be used at any one of a number of points including; design verification, component verification, assembly verification, test, and incoming inspection. In all cases, the goal is to precipitate any latent failures of electronic components or products.

Statistically, the pattern of failures in product follow a classical bathtub curve (Figure 16-1). Once the statistical anomalies have been removed, the physical process closely approximates the curve of the idea model. The failures of components early in the design lifetime are due to two primary factors: weak components and improper manufacture. Weak components are statistically predictable, and they in turn have bathtub curves associated with them. The so called learning curve (ideally an inverse exponential curve) contributes the remainder of the slope to the initial part of the bathtub curve.

Ignoring overstress failures, random failures are generally evenly distributed during the bulk of a product's lifetime, producing the horizontal portion of the curve. Only when the statistical design life of individual components is reached does the curve begin increasing. In our discussion, we will refer to the three parts of the curve as early failures, normal lifetime, and age failures.

The term burn-in is a misnomer. The process step known as burn-in may consist of operation at artificially elevated (or depressed) temperature, or with marginal cooling. The test may be performed with high or low input

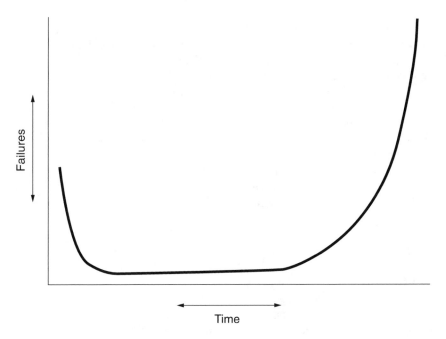

Figure 16-1. Bath-tub Curve

voltage, and in some cases other environmental tests (such as humidity and radiation) may be performed. In some cases operation is performed in typical conditions. In other cases environmental stress is applied before (but not during) electrical operation.

Burn-in steps may take a few minutes per unit, or might conceivably require months to complete. The length of time spent performing burn-in is a function of the type of units being tested, the reliability required, the history of the particular burn-in process, the susceptibility of the units being tested, the projected failure modes, the design lifetime, and the number of parameters being evaluated. Other factors may also affect burn-in steps, such as product volumes, burn-in station throughput, and complexity of test. More likely than not it is the complex interaction of these factors that determine both the time a product spends in burn-in and the effectiveness gained.

Measurements performed during or after burn-in determine the reliability of the units tested. These measurements range from simply detecting the presence of expected outputs to intricate parametric analyses of transfer functions and device action. The types of measurements performed are determined by the types of units under test, the failure modes of concern, the precision of expected results, the accuracy of measurement tools, the costs involved, the design tolerances, and importance of parameters, among other factors.

It is possible to use the test to statistically determine the shape and length of the initial part of the bathtub curve. Some situations are extended to life-testing, where the intent is to find the statistical life expectancy of a particular type of component or assembly, and thus the shape and starting point of the final part of the curve. In addition, burn-in may be used to accelerate early component failures, so that survivors may be reasonably expected to be in the flat portion of their life cycle curve. In the statistical instances, sampling is done to provide enough data for analysis. In the final case, 100% testing is required to adequately reduce early component failures. The implementation of burn-in for statistical purposes may be a temporary process step which is eliminated when the statistical maturity of a process and the components is assured.

Quality

Six Sigma

The Six Sigma program sets the quality for engineering products, and controls manufacturing processes in order to approach the goal of zero defect performance levels. "Sigma is a statistical unit of measurement that describes the distribution about the mean of (the results of) any process or procedure. A process or procedure that can achieve plus or minus Six Sigma capability can be expected to have a defect rate of no more than a few parts per million, even allowing for some shift in the mean. In statistical terms this approaches zero defects. Motorola's goal is to achieve this level of quality in everything that Motorola does." The preceding is an excerpt from a brochure available from Motorola titled *Quest for Excellence*, and can be obtained by calling Motorola University at 1-847-538-2568.

Another resource on the Six Sigma program is consultant Ron Minezzi.[12] Mr. Minezzi was trained by Motorola as a Six Sigma Trainer while he was a manager in the corporate quality group at Digital Equipment Corporation.

Cost of Third Party Repair

The cost of repair must include, not only the actual cost of repair, but also:

- The cost of transportation, including paperwork processing for transportation, and any insurance costs, unless you are self-insured, in which case your insurance department can assist you in calculating the correct insurance costs.

- The cost of ownership of the defective units from the time you receive them until you receive them back from repair and put them into inventory. This is sometimes called pipeline cost.

- The cost of additional units to maintain inventory levels due to filling the pipeline.

- The loss of timely feedback from repair to engineering can drastically increase costs until an engineering change can be made to a product. Worse yet is the fact that if a single problem is accounting for much of the work with a particular type power supply, the vendor can quickly diagnose it and repair it with a better profit margin, and may therefore be somewhat reluctant to quickly notify you of the design flaw. Or the vendor may simply fail to recognize it at all. This can quickly become very expensive, but worst of all is the loss of goodwill with your customers who continue to have failures, of which you are not yet aware, that constitute a serious problem with the design or with the quality of a part or parts in the product.

- The loss of control of product repair cycles can occasionally be a problem. When the repair is inhouse, and you need a quantity of Model X before Model Y, you can easily and quickly set priorities. This usually

---

[12] Mr. Ron Minezzi, 134 Center Road, Shirley, Mass. 01464 (508) 425-4953.

requires extra effort and planning to achieve with an external vendor. This fact must be considered from the start of negotiations to create the right business relationship with the vendor. One way to avoid this by making use of just in time (JIT) as at least one repair house has.[13]

- The cost of material may be somewhat higher for the small repair vendor who purchases replacement parts at a much lower volume than your company. This is especially true for parts that are still used in volume production, or part of a blanket order that covers many parts at higher-volume levels.

- The cost of any taxes that may be imposed, especially value added taxes (VAT). Taxation is common when shipping product to another country for repair.

Feedback from Repair Operations

Perhaps one of the most important considerations, which is usually forgotten in negotiations for a repair contract, is the aspect of feedback of actual repair data. If the data is not useful and timely and is not in a usable format it may be of little use. To say that 12 resistors, were replaced last month is of no use. The data must include the model number, serial number, the fault, the faulty component, and why it was faulty. When the number of parts failing for a part number exceeds a preset level, all the defective components will be returned for examination or examination by the company who supplied them originally.

The presence of predetermined failure levels for each component type will flag the repair vendor to immediately notify the appropriate contact so that he or she will quickly act on any needed engineering change for either manufacturing, repair, or both.

Dangers of Third Party Repair

You must also consider the possible dangers of third-party repair, other than those that have already been mentioned. One such danger worth mentioning is the stability of the vendor company and its management team. The smaller the company, the larger will be the perception of risk.

---

[13] Bruce Rayner, "Zealous Zytec Takes on the World with JIT," *Electronic Business*, November 1, 1987.

Annotated Bibliography

1. ———, "Quest for Excellence," Motorola Inc, Motorola University, Ph 1-847-538-2568.
2. Bormann, Joan, "Hands on JIT, Zytec Corporation's Pains and Gains," *The Inside Line,* April 1988.
3. Hildegrant, Earl, "Supply Management in Minnesota," Target Association for Manufacturing Excellence, Inc. Spring, 1989.
4. Jensen, Finn; Petersen, Neils Erik; *Burn-in, An Engineering Approach to the Design and Analysis of Burn-in Procedures,* Wiley 1983.

    This book analyzes component and assembly burn-in from the statistical view. Mathematical methods and models are given to explore the fundamentals of burn-in statistics. Throughout the book attempts are made to relate the statistics to the real life problems of production engineers for example, how to determine the amount of burn-in required to achieve a particular level of confidence.

    The comparison of burn-in with and without repair is made. Methods of predicting reliability and effectiveness of the burn-in process are explained. The differences between the "infant" failure and "freak" failure are covered.

    Much of the analysis in this book is based upon previously published material by other experts in the reliability and statistical analysis fields. Without referencing those works, understanding the derivation of certain formulas is difficult. More than 150 publications are referenced in the bibliography. Moreover, the authors freely use statistical concepts which are not commonly encountered by the common production engineer.

    The book is perhaps most useful as a reference guide, and not as a handbook. The authors cover the material in depth sufficient to allow the reliability engineer (their intended audience) to make judgments about proposed and existing manufacturing processes, and the effect of changes on those processes.
5. Karam, Douglas, "Burn-in: Which Environmental Stress Screens Should Be Used" Griffiss Air Force Base, New York 13441, Rome Air Development Center, Air Force Systems Command, March 1981

    This 61 page report is based on existing literature, and concludes that thermal cycling and random vibration are the two most effective screening methods. This report was generated to serve as a recommendation for a military burn-in standard. The models examined by Hughes Aircraft Company, IBM, NAVMAT, and Martin Marietta are discussed in this report. Of particular interest are the empirical screening formulas which evaluate screening effectiveness based upon time, temperature, vibration, number of cycles, and type of stress. These formulae are derived in a report by Hughes, which is listed in the bibliography of the Karam report. The report graphically represents the effects of lengthened test time, elevated temperature, and other parameters. The types of failures precipitated, and the threshold of diminishing returns are explained.
6. Mankikar, Mohan, "Power Technologies Pursuing the Perception of Quality," *Electronic Engineering Times,* June 11, 1990.

7. Murray, Tom, "Just-In-Time Isn't Just for Show It Sells," *Sales and Marketing Management,* May 1990.

8. Saari, A.E.; Schafer, R.E.; VanDenBerg, S.J.; "Stress Screening of Electronic Hardware," Hughes Aircraft Company, Fullerton, Ca 92634, February 1982.
    This 183 page report recapitulates the findings of earlier reports (from Hughes Aircraft and others), concerning the use of thermal and vibrational stress screening. The parameters used in several previous studies are compared. Discussion of cost savings as compared to screening program cost is included. The report includes runs from reliability analysis programs using the Stress Screening Model (SSM) which is a variation of the Screening and Debugging Optimization model (SDO). Pertinent formulas are listed in the Appendices.

9. Taylor, David, "Faxban: Kanban Meets the Fax Machine," *Target,* Summer, 1990.

10. Taylor, David, "JIT and the Fax Machine," *Purchasing,* September 13, 1990.

17

Repair Options

Repair Site Options

There are several options for the repair of power supplies that fail in customer products. The choices include:

- On-Site Repair
- Regional Office Testing and Repair
- Field Service Depot Testing and Repair
- Factory Testing and Repair
- Third Party Repair Operation
- Scrap and Replace
- Combinations

In this chapter we will examine these options in more detail.

On-Site Repair

On-site repair can be the only choice in repair of very large power supplies such as extremely high power units of either high voltage or high currents. Large units, for example, a 1.2 Megawatt power supply for a Liner Accelerator would usually best repaired on site. The factors to be considered include available spares, shipping time, shipping costs, downtime considerations, portability, form factor, ease of removal, available test equipment, available test loads, and source power availability. When the situation demands it, or when it is the most cost-justifiable solution, on-site repair may become the most logical choice. A properly trained and prepared technician or field engineer working with the customer's site personnel can effect the repair and test the power supply in the system after the repair work in completed.

On-site repair cannot be the choice in most cases, however. Smaller high-volume products cannot cost effectively support the on site repair effort. Swapping a bad power supply for a good one is the most logical choice. On-site repair not being the choice indicated that the repair is performed elsewhere in the supply chain. On-site repair is quite often achieved by module replacement within the power supply.

In larger systems there may be several power supplies, or there may be power supplies containing several modules. Repair may be accomplished by replacing the defective module or modules. Where some level of built-in test has been built in, there may be some indication of which modules or power supplies are faulty. The level of built in test is usually limited. There are exceptions, some with very elegant built-in self-test provisions. As more and more smart power controllers are used in power systems, built-in self-test will become more commonplace. Although still rare, remote diagnostics can be achieved in power systems. It is usually employed in systems where high maintainability is required or where physical access to the power supply is not available. To reiterate, smart power controllers will eventually increase the usage of remote diagnostics for larger power supplies in the future.

Regional Office Testing and Repair

Regional office testing and repair is the logical cost-effective choice under some of the following circumstances:

- High repair volumes
- High taxes
- Long shipping times
- High no problem found (NPF)
- Lower local repair costs
- Spares availability

High Repair Volumes When high enough volumes of a particular power supply occur, it may become more cost effective to repair the defective power supplies locally. The decision must be driven by the total cost of repair to the corporation, for all power products, and not just on local costs alone. Consideration must also be given to the other repair alternatives and what negative impact may occur due to changes in volume from diverted local repair operations. For many, the question of the control of quality and product revision due to ECOs may preclude local repair efforts. In some instances local repair operations, driven by the need to satisfy local customers and cost constraints, may actually take place without corporate sanction. Should this problem occur, it needs to be addressed and solved by all those involved, at both local and corporate levels.

High Taxes   In some countries on this planet, extremely high value-added taxes (VAT) or other import/export taxes may preclude shipment out of the local country for repair purposes. In this case local repair is a solution to staying both competitive and solvent. In one instance the value added tax exceeded the original cost of the power supply by 200%.

Long Shipping Times   In some instances, a long shipping time which increases the actual cost of repair by requiring higher spares levels, and the associated cost of ownership of the power supplies in the "pipeline" drive the cost of repair out of the competitive range. Local repair then becomes an option that otherwise may not have been pursued.

High No Problem Found (NPF)   In the case where certain power supply products exhibit a high no problem found (NPF) rate, local test and perhaps repair of some defects, becomes a logical choice. The solution to the problem of many NPF in power supplies can usually be found in good power supply design, packaging, and applications.

Lower Local Repair Costs   When the local cost of repair is very low as compared to factory or depot repair, power supplies may be repaired locally as a cost cutting option. It is important to remember that the actual costs both for the local repair and the "cost" associated with not using the corporate repair option make up the total cost of local repair for a product.

Spares Availability   When spare power supplies are unavailable and repair cycle time is critical, local repair may be the only cost-effective choice available. Lack of spares can occur on both brand new products and products at or beyond end of life where all or most of the spares for that product have been consumed.

Field Service Depot Testing and Repair   The creation of a field service depot for testing and repair of power supplies may be the logical choice in very-high-volume operations where the cost of a service depot can be justified. In an operation where a large variety of power supply repair is required with any one type providing a very small number of failures a depot operation can be very cost effective. One example of this type of operation is the military depot where low volumes of any individual power supply type add up to a large volume of repair. Large third party repair operations may be similar in nature.

Factory Testing and Repair   As a new products volume in creases, cycling field failures back through the manufacturing plant will provide feedback to both engineering and manufacturing on those field returns. Once the manufacturing volume has reached its peak, some operations maintain the repair operation for field returns through end of manufacturing life. Some manufacturers will even maintain the repair operation through the end of product life. Some manufacturing operations will transfer this field repair process to a field depot as soon as the depot can put the process in place. The different philosophies of manufacturing are beyond the scope of this document, however, each has its advantages and disadvantages. The choice is really up to plant and corporate management.

Third-Party Repair Operation   The use of a third-party repair facility for testing and repair of power supplies may be the logical choice for some. While there are a number of concerns that must be addressed, a properly managed relationship with a third-party repair vendor may be a cost-justifiable option worth consideration. A number of vendors are in this business. One of the larger of these vendors is ZYTEC in Eden Prairie, Mi. You can contact Doug Franzen at Phone 612-941-1100 ext-142.

Scrap and Replace   For some power supply products, the most cost effective repair option is to scrap and replace. These include certain low-cost units, encapsulated units, small hermetically sealed units, and units packaged such that repair is not an option. This decision must be carefully made due to cost impacts, and the loss of feedback to engineering on failures. A full financial analysis must be made to make the proper choice in each case. When scrap and replace is used, a decision must be made as to how the scrapped units will be disposed of. It may be wise to have them crushed for metal salvage to prevent numbers of defective units from becoming available on the low cost surplus market.

Combinations   In all reality, it is usually a combination of these options that together make up the best solution for any one company or division. Examining all your options with all of the parties involved and arriving at a consensus will help ensure that the best solution win be found and implemented smoothly.

Repair Options   There are three options for selecting the repair strategy for a product. They are:

- Test only, no repair.
- Test and repair some.
- Test and repair all.

Test Only   Test only for the purpose of determining the disposition of each power supply may make sense to prevent the shipping of unit that will be scrapped. Bulk shipping of scrapped units may make sense if the power supplies are received by regional centers or a corporate center and shipped elsewhere for repair. This may especially be true for both high voltage power supplies and power supplies built with surface mount components due to the diagnostic and repair capability needed for handling these types of power supplies.

Test and Repair Some   Testing and repairing those that can be completed quickly without excessive diagnose time or material costs, can be a very cost-effective option. Those products that exceed a predetermined time limit for diagnose or parts cost for repair may not be cost effectively repaired. For new product start-up, determining the cause of failure as a source of data to repair or adjust the process, the immediate cost is less of a consideration. Testing sealed or potted units prior to scrapping or shipment to a repair facility can also make sense. The biggest danger here is that the tests performed must be a burn-in type test so that thermal failures and units that drift out of specification can be detected and appropriately dealt with.

Test and Repair All   While this is the stance that most manufacturers take, it is not always the correct one. At some point, the cost of repair may exceed the cost of replacement. A few troublesome units can drive the average cost to repair and the mean time to repair higher than desirable.

Test Equipment

In all situations, capital equipment cost are a concern. In the small operation or a branch office one solution that may prove very useful is the technique of nodal impedance signature analysis. This technique allows a nonpowered test of two boards for comparison and determining which parts are in need of replacement. While this technique is not a substitute for test, it is a method of diagnosis that can be easily accomplished with minimum skill level repair technicians. Nodal Impedance Signature Analysis equipment is available from several vendors including:

- Searcher by ATS Inc, Middleton, MA.
- Huntron Tracker by HUNTRON, INC., 15720 Mill Creek Blvd., Mill Creek, WA 98012 (206) 743-3171 or 1-800-426-9265

No Problem Founds and Intermittent Problems

One very serious problem that can occur is the return to service of power supply units or boards where no problem has been found, or where the problem is of an intermittent nature and has not shown itself. It is recommended that units falling into this category be given the special attention of an extended test and a thorough burn-in, or environmentally controlled stress test to bring the unit under test to full operating temperatures. This will help to prevent intermittent problems from being inserted into customers equipment and/or systems.

18

Depot Testing and Repair

Choosing Products

For each product under consideration a number of factors must be considered, and decisions made. These decisions include repair or scrap, where to repair, when to repair, what process to use, and how to repair. Each product must be considered individually and as a part of the whole repair strategy.

Repair Depot Test Processes

Typically, depot repair means a high-volume of units, many of the same type. This volume comes from the large number of units in the field that have been sold to customers. These high volume levels mean that the processes that are used will be similar to the high-volume manufacturing processes. Some special considerations should be observed, however.

Special Tests for Field Returns

In addition to the factory type of tests, some special tests are required for field returns to indicate reduced meat time between failure (MTBF) or life values due to the aging of some components. For power supplies, the single major component that requires special consideration on all returned field units is the electrolytic capacitor. Electrolytic capacitors age, due mainly to the drying out of the electrolyte, a liquid. There are two tests that may be used to identify problems which may not show up under other test conditions. These tests are:

1. Efficiency
2. Ripple at low-line full-load conditions

Efficiency

Efficiency is a measure of the power consumed in the process of power conversion. For DC-DC converters, it is simply the Watts Out divided by the Watts In, multiplied by 100%. For AC-DC and AC-AC converters, the measurements process requires that careful attention be paid to the RMS measurement of both the voltage and current, as well as the phase relationship between them.

For some types of switching power supplies, a good indication of losses can be seen by shorting the output(s) and measuring the input power.

Ripple at Low-Line and Full Load Conditions

The input capacitor will generally age fastest due to the higher voltages, causing higher leakage currents and more heating than found in lower-voltage output capacitors. It is therefore appropriate to perform some testing on units returned from the field to determine that they will not fail soon after repair due to poor electrolytic filter capacitors.

By performing a full load test at some specified voltage just below the low line voltage specification, a determination of the point at which AC line ripple bleeds through to the output(s) can be made. This test, somewhat crude in nature, can be used to indicate the relative value of the capacitor(s). Then the decision can be made whether the input filter capacitor(s) needs to be changed to prevent an unhappy customer in the very near future.

Repair Depot Test Tools

Depot test tools may include in-circuit testers (ICTs), power supply testers, and nodal impedance signature analysis instruments.

In-Circuit Testers

As the volumes of particular printed circuit boards become sufficiently high to justify the cost of an in-circuit tester, the machine becomes the diagnostic element in the process when properly programmed. As discussed above, an in-circuit test system like the GenRad Mixed Signal In-Circuit Tester manufactured in Concord, MA is the author's choice. This tester can perform the required level of in-circuit testing and loop testing to give the level of coverage required in the power supply boards being tested. Board coverages of 99%+ are not uncommon when using these techniques. This removes the costly diagnostic technician from the loop for most if not all of the products being tested and repaired.

Unit Testers

In smaller operations, less expensive power supply test systems will have a more prominent role. Otherwise, testers available off the shelf can meet the need of test. When possible, it is desirable to use the same tester that manufacturing uses when testing new power supplies. As production volumes decrease for a particular product in manufacturing, dedicated testers may become available to repair depots. This can become a very costly proposition as compared to off the shelf systems that can be reconfigured and programmed to meet a wide range of products as volumes dictate.

Nodal Impedance Signature Analysis

Nodal impedance signature analysis is a low cost method of quickly diagnosing printed circuit boards at low volumes without using highly skilled technicians. It is a comparison technique where a known-good board is compared, node by node, with the suspected bad board. This technique may have some value in smaller operations where volumes are just not high enough to justify the cost of an in-circuit tester (ICT).

Two vendors of this equipment are:

- Huntron Tracker by Huntron Inc, 15720 Mill Creek Blvd., Mill Creek, WA 98012 Phone 206-743-3171 or 1-800-426-9265
- Searcher by ATS Inc, Middleton, MA.

Either manufacturer will gladly provide information on their products.

Factory Repair

Some companies opt to perform factory repair where the repair facility is located at the factory either in a separate department, or integrated into manufacturing operations.

Separate versus Integrated Repair Operations

You will have to decide which option is best for your company, and either segregate or integrate the field return repair operation as your needs dictate.

Again, take the global view from the corporate position, and work with those who are involved in any way with the operation from both manufacturing and field service.

Cost of Repair Operations

Working with your team of experts from all departments, you will have to determine the cost of repair for each product.

Feedback from Repair Operations

One of the most important, and most often overlooked source of information on the quality of the product being shipped, comes from the repair operation. At one large manufacturer of electronic equipment containing power supplies, a bag of a thousand diodes of the wrong speed but with the correct appearance, found their way into hundreds of power supplies. The incorrect diode was used in a snubber circuit and the only failure mode was a greatly shorted MTBF; in fact, shortened to about 40 hr, instead of a number thousands of times higher. The repair operation saw close to a thousand returns cycle through the repair operation in only a 3 or 4 month time frame. By the time that anyone noticed that the volume for this particular product was ramping up to excessive volume, and discovered the problem, only a few of the offending diodes were still in the incorrect box in the stockroom. The rest were in product. This was an expensive and painful lesson.

Testing Internals

The only tests we will examine here are ones where the test or measurement is made for a different reason on field returns than on new units in manufacturing.

Output Backpower Test

Check all outputs for shorts prior to power being applied to the product. This is important because most field returns are bad by definition. In manufacturing the opposite assumption is made.

Reference Voltage Checks

Check all reference voltages prior to power application. This is important because most field returns contain components that have aged, and for critical components like voltage references, must be checked.

Softstart Test

This test is designed to check the softstart functionally prior to power being applied to the supply. This can be important where input power components may have been, or may be, damaged.

Main Switch Transistor Short Test

This is a test to check that a short doesn't exist on main switching transistors prior to power being applied to the supply. This is important for field returns due to the fact that many failures are in the main switching transistors.

Input Current Ramp Up Test at Minimum Load

Check that an AC short doesn't exist on the supply prior to power being applied.

High Line Minimum Load Regulation Test

This test to check regulation at high-line minimum-load is important or field returns because it applies the maximum voltage stressing on the input power circuitry.

AC Input Current (RMS)

Checking the AC RMS input current may be important looking for poor efficiency and high losses.

Low Line Full Load Regulation Test

The regulation at low-line full-load test is usually done on the 115 V switch position to worst case the input current. This is perhaps the most important test for field returns. Tired and/or weak input electrolytic filter capacitors will cause excessive ripple during this test.

PARD or Periodic and Random Deviation and Ripple Test

The periodic and random deviation (PARD) and the ripple tests are made to check the outputs of the supply for excessive ripple and noise. Output filter capacitors can cause problems that are detected here. This test should be made at the same performance levels as the previous test.

Dynamic Loading Tests

This test is targeted at checking the closed-loop response of the regulation amplifiers. For field returns, the test seeks to identify any response problems due to other small electrolytic capacitors in the amplifier circuitry.

No Problem Found Units

It is of paramount importance that any unit or module that cannot be diagnosed as having a problem, be diverted for special testing for intermittent failures. This test should consist of at least cold soaking, an extended environmentally controlled stress testing (burn-in), and perhaps even a test on a shake table. Shipping intermittent failures to the field is a sure way to destroy a repair depots reputation.

Index